Michael Fuchs, Maria-Theresia Holub (eds.)
Placing America

American Studies | Volume 3

Michael Fuchs, Maria-Theresia Holub (eds.)
Placing America
American Culture and its Spaces

[transcript]

Printing costs funded by the University of Graz, the City of Graz, and the Unit on Science and Research of the Province of Styria.

Bibliographic information published by the Deutsche Nationalbibliothek
The Deutsche Nationalbibliothek lists this publication in the Deutsche Nationalbibliografie; detailed bibliographic data are available in the Internet at http://dnb.d-nb.de

© 2013 transcript Verlag, Bielefeld

All rights reserved. No part of this book may be reprinted or reproduced or utilized in any form or by any electronic, mechanical, or other means, now known or hereafter invented, including photocopying and recording, or in any information storage or retrieval system, without permission in writing from the publisher.

Cover layout: Kordula Röckenhaus, Bielefeld
Cover illustration based on Gabriel Millos' photo »Route 66«, published under
   Creative Commons 2.0 licensing on the French Wikipedia on September 26, 2009; edited by Michael Fuchs.
Proofread by Michael Fuchs/Maria-Theresia Holub
Typeset by Michael Fuchs
Printed by Majuskel Medienproduktion GmbH, Wetzlar
ISBN 978-3-8376-2080-1

Contents

Placing America: Constructing America through Time and Space
Maria-Theresia Holub | 9

CONSTRUCTING AMERICA FROM AFAR

Performing America Abroad:
No Name City and the Haunted Spaces of Transnational America
Leopold Lippert | 19

America, the Threat of Time:
Sigmund Skard and Early American Studies
Ida Jahr | 39

REAL PLACES AND IMAGINARY SPACES

Setting the Scene:
L. M. Montgomery's Imaginative Island Landscapes
Julia van Lill | 57

Fallujah Manhattan Transfer:
The Sectarian Dystopia of Brian Wood's *DMZ*
Georg Drennig | 75

There's No Place Like Fiction: Narrative Space and
Metalepsis in Stephen King's »Umney's Last Case«
Jeff Thoss | 91

The Black Hole at the Heart of America?
Space, Family, and the Black Hallway in *House of Leaves*
Michael Fuchs | 103

DRAWING BORDERS

Meeting at the Border:
The Canadian ›Two Solitudes‹ in Érik Canuel's *Bon Cop, Bad Cop*
Yvonne Völkl | 129

›Romanized Gauls‹: The Significance of the United States and the Canada-U.S. Border for Canadian National Identity Construction
Evelyn P. Mayer | 145

MARGINALIZED CULTURAL SPACES

The Fine Line between Utopia and Dystopia:
Representing America in Thomas Pynchon's *Mason & Dixon*
Diana Benea | 161

Getting a Name: Searching for a
Mixed-Blood Identity in Sherman Alexie's *Flight*
Madalina Prodan | 173

This Space Called Science: Spatial Approaches, Border Negotiations, and the Revision of Cultural Maps in Contemporary Popular Culture
Judith Kohlenberger | 187

Contributors | 205

Index | 209

Acknowledgements

This book evolved out of a conference entitled »Space, Place, and Time: The Construction of Identity in American Literature and Popular Culture«, held in Graz, Austria, in December 2010. Thus, our first thanks go to all the conference participants, among them also those whose essays we weren't able to consider for this collection: Thank you all for your thoughts and contributions to the discussions, a number of which are reflected in this volume's chapters.

Both the University of Graz and the City of Graz offered their generous support to the event, for which we are very grateful. To Professor Walter Hölbling, who provided introductory remarks and tirelessly participated in discussions during the conference: Thank you for your presence and for your critical inquiries! Since you recently retired, we wish you all the best for this new stage in your life!

Last, but not least, we would like to acknowledge the financial support for this volume by the University of Graz, the Province of Styria, and the City of Graz, without which this publication might not ever have seen the light of day.

Placing America
Constructing America through Time and Space

Maria-Theresia Holub

> If history is not only temporal or chronological, but also spatial and relational (and if, conversely, our understanding of geography itself is never historically innocent) then it follows that our analysis of ideas of postmodernity must consequently be informed by this kind of geohistorical perspective.
> David Morley, »EurAm, Modernity, Reason and Alterity«,1996

Explaining why she would not vote for Barack Obama in the 2012 Presidential election, Tea Party Express Chair Amy Kremer claimed that »[for Obama] it's not about the Shining City on a Hill, the greatness that has always been America, that our Founding Fathers were about« (qtd. in CNN 2012: par. 176). Although Kremer's insistence on American exceptionalism and its Puritan founding myth may be naïve, even questionable, it still points to some relevant issues, namely (among others) the enduring centrality of Puritan spatial metaphors within mainstream American discourse and the idea that the 'foundation' of the United States marks both a specific moment and place in time.[1] While she may not harbor much interest in or appreciation for a critical American historiography, Kremer still unwittingly confirms David Morley's claim quoted at the beginning of this introduction that space (in Kremer's understanding: the image of the ›City Upon a Hill‹) and time (the History of the preeminence of American leadership) are to be considered mutually constitutive and that one cannot be adequately analyzed without paying attention to the other. Morley discusses space and time in the context of postmodernity, which adds an interesting twist to the discussion. For, not unlike the rhetoric of the City Upon a Hill or the related American Dream, postmodernity posits itself as a democratizing force, questioning hierarchies while simultaneously obscuring its own perpetuation of imperialist structures.[2]

In *Culture and Imperialism*, Edward Said similarly links history and space when he contends that »[t]he appropriation of history, the historicization of the past, the narrativization of society [...] include the accumulation and differentiation of social space« (1993: 78). By offering a more nuanced understanding of space historically (and of history spatially), Morley's proposal of what he calls a ›geohistorical perspective‹ seeks to make visible the construction behind such naturalized narratives evoked by Said. The question that emerges is: How would such a geohistorical perspective on America (or more specifically, the United States) look like? What are the physical and epistemological demarcations of space in America? Or, in other words: When and where can we locate America?

One might start by probing the significance of spatial metaphors in the construction of a historical identity of the nation. What images are needed to conceptualize America? The aforementioned ›City Upon a Hill‹ certainly is one of the earliest and most enduring ones. It was most prominently employed by John Winthrop who, using a verse from the Bible, reminded the Puritan brethren that »wee must Consider that wee shall be as a Citty upon a Hill, the eyes of all people are uppon us« (1630/1994: 233). The image became so central to the country's »national imaginings« (Anderson 1983/1991: 9) that numerous U.S. politicians of both major parties have utilized Winthrop's phrase to market their agenda of American exceptionalism, a myth that, as William Spanos explains, »has determined not only the unilaterality of American foreign policy from the beginning of its existence but the very American culture on which this benignly aggressive foreign policy depends for its practice« (2003: 30). A speech given by John F. Kennedy provides a famous instance in which he likens his presidency to the first Puritans in the ›New World‹, faced with »the task of building a new government on a perilous frontier« (1961/2012: par. 12). The ›Shining City Upon a Hill‹ is then not to be mistaken for a peaceful parish. It rather marks the construction of an imperialist center set on expanding its territory. At the time Kennedy delivered his address, the Cold War was well underway, and the competition between the Soviet Union and the United States for preeminent leadership extended from global territory to outer space. Finally, in their most recent manifestation, American notions of expansionism and manifest destiny have transitioned from a war for territory to a war on terror.

While idealized notions of early Puritans ›roaming in the wilderness‹ abound, it becomes quite clear that space, in American history, has hardly ever been innocent and has very frequently been tied to imperialist agendas. When Klaus Benesch refers to the idea of space in America as »a contested terrain, a site of continuous social and cultural struggle« (2005: 19), he does, of course, not solely mean exterior sites of conflict, for the contestation of land has been an inherent part of American identity since the country's inception. The apparent wilderness encountered by the Puritans was, in fact, heavily populated by a number of nations with different languages and customs, who ended up being decimated and relocated to reservations, because they somehow wouldn't want to accept that they would never be considered anything beyond the primitive heathens the Puritans so desperately sought to label them. As Jean Baudrillard claims in his reflections on America, »[h]ere in the most moral society there is, space is truly immoral« (1986/1999: 9). It is certainly never innocent or ›free‹, as, for instance, Frederick Jackson Turner implies in his essay »The Significance of the Frontier in American History«, first presented to the public as a lecture in 1893.

Only in the twentieth century has there finally been some acknowledgment that the acquisition of territory did not take place in an ›uncivilized‹ void, and that the repercussions of the conquest of America can still be felt today, not only by Native Americans, but also by African Americans, whose very presence is a reminder of a trauma underlying American society, one that is as much tied to territory as it is to race. When Martin Luther King, Jr. delivered his famous »I Have a Dream« speech in Washington in 1963, he disclosed the American Dream of equality as a mostly unattainable myth, especially to those outside the white, middle-class mainstream. As Howard Zinn explains, racism and notions of racial superiority became institutionalized, among other reasons, to diminish the chances of crossracial solidarity amongst servants and slaves. Constructed hierarchies are thus presented as a narrative of nature and biology, when, in fact, racial markers ensuring hierarchical opposition are »historical, not ›natural‹« (1980/2005: 38). American history is thus just as much a story of the invention and expansion of space, as of its regulation and confinement.

If membership in the ›imagined community‹ of the nation is also contingent upon factors such as race or language, what about those who cannot claim one distinct side for themselves? Gloria Anzaldúa was among the first to give voice to those socially and politically displaced in the borderlands, »a vague and undetermined place created by the emotional residue of an unnatural boundary« which appears »in a constant state of transition« (1987/1999: 25). Rather than the much appraised ›threshold of opportunity‹, the borderlands Anzaldúa delineates are spaces where transgression is both imminent and dangerous. The ›mestiza consciousness‹ Anzaldúa conceptualizes closely links identity to the spatial markers of geographical, cultural, social, spiritual, and psychological borderlands, the spaces between and beyond conventional dichotomies. The frontier, hailed by Turner as a symbol of progress and democracy, is challenged by the Mexican–American frontera, highlighting how borders may serve both as bridges and boundaries.

While Border Studies has for quite some time been mainly associated with the Mexican–American frontera and with Chicano identity, in recent years the study of the U.S.–Canadian border has finally gained more scholarly attention. There is finally more of an understanding that ›America‹ extends beyond the borders of the U.S.-American nation-state. Hence the essays in this volume not only present space in general as a multi-faceted field, but they also contend that issues of space in America pertain just as much to Canada as to the United States.

Although Canada's status as a ›Western‹ country makes it apparently less necessary to ward off unwanted illegal immigrants through American security border fences, the relationship between the two countries still remains unequal, with Canada positioned as a junior partner to the United States. In Canadian writer Margaret Atwood's words, »Canada is an odd country: patriotism has always been regarded with some suspicion in it, because—as in any satrapy—getting too uppity about yourself might offend the imperial centre and thus be bad for business« (1972/2012: xxiv). The ›imperial centre‹ here refers to Canada's southern neighbor. Yet Canada has had its share of conflictuous relationships within its own boundaries, too. They include, for instance, the animosity between the French-speaking Quebecois and the English-speaking rest of the country and, on a different level, the ongoing marginalization of First Nations people who, although not decimated as violently as in the United States, have repeatedly found themselves on the weaker end in their relations with (the mostly white) mainstream Canadian society.

Despite the intricate ties to issues of dominance and oppression in the instances of space depicted above, I in no way want to suggest that spatial relations in the Americas are solely to be explained in this manner. However, as conceptualizations of space »are tied to the relations of production and to the ›order‹ which those relations impose, and hence to knowledge, to signs, to codes, and to ›frontal‹ relations« (Lefebvre 1974/2000: 33), a more nuanced discussion of space in America also needs to pay attention to such factors. In the present collection of essays, we have tried to be mindful of these codes, signs, and relations.

This book is not intended as a definite take on the topic at hand. Instead, it seeks to take up discussions already existing in the field and engage them critically. In the introduction to *Thirdspace*, Edward Soja proclaims that it »becomes more urgent than ever to keep our contemporary consciousness of spatiality—our critical geographical imagination—creatively open to redefinition and expansion in new directions, and to resist any attempt to narrow or confine its scope« (1996: 2). The aim this book is precisely that: to provide a multiplicity of voices and to feature various conceptualizations of space, be they geographical, social, or epistemological, and to place them in dialogue with one another.

The first section of the book at hand focuses on conceptualizations of American space from afar. Taking the movie *No Name City* by Austrian director Florian Flicker as his point of departure, Leopold Lippert discusses Americanness as a deterritorialized specter conjured up in the Austrian Western city ›No Name

City‹ near Wöllersdorf. Rather than viewing the nation-state and national identity as naturally given, Lippert proposes American identity as an act of global performance. Ida Jahr's »America, the Threat of Time« zeroes in on the Norwegian Americanist Sigmund Skard who, while being a European pioneer in American Studies, developed an ambivalent position toward America: to him, American culture became equally a symbol of progress and modernity as well as a potential destabilizer of the Norwegian nation-building project. Jahr traces the tensions between these two perspectives, chronicling Skard's America as a heterotopic space of ambivalence.

Placing America's second section moves on to the realm of the imaginary. First, Julia van Lill probes the imagined landscapes to be found in Lucy Maud Montgomery's young adolescent fiction. For instance, in her novel *Anne of Green Gables*, space is constructed and shaped by the protagonist, the young orphan girl Anne, who finds new names and identities for the various natural sites in her surroundings. Place thus functions as a signifier of the young girl's identity and creativity. More specifically, the exterior natural landscape comes to be linked directly with the protagonist's interior space of imagination. Following her essay, Georg Drennig's »Fallujah Manhattan Transfer« looks at the comic book series *DMZ* and its futuristic depiction of Manhattan as a divided city, a war zone where various groups fight equally for survival and control, each laying claim to representing the ›real‹ America. Using Foucault, Drennig analyzes the inner-city war zone as a dystopian space in which those elements signifying an authentic American sense of identity ultimately cause the city's demise. In his chapter, Jeff Thoss takes us to the border between fiction and reality. Through the concept of metalepsis, that is, transgressions between narrative levels, Thoss analyzes the ways in which various spaces are unsettled and rearranged in Stephen King's short story »Umney's Last Case«. Ultimately, Thoss argues, metalepsis becomes not an indicator of the interrelationship between reality and fiction. Instead, it signifies a shift toward a re-definition of the border between these two ontological levels that seems to have dissolved in postmodernity. Michael Fuchs' »The Black Hole at the Heart of America?« discusses a text that eludes easy categorization: Oscillating somewhere between a novel, an academic treatise and the written representation of a documentary movie, *House of Leaves* challenges the reader's perception through various layers and modes of narrative, thereby creating a complex web of stories and histories, at the heart of which lies a typical American family. Reading the text(s) as a haunted house tale, Fuchs critically assesses quintessential American themes, such as home or family.

The third section of the book moves from the realm of the imaginary into actual border spaces within and around the Canadian nation-state. Yvonne Völkl's essay focuses on the real and imagined borders between the French-speaking Quebecois and the English-speaking part of the country. Völkl's analysis of the

Canadian movie *Bon Cop, Bad Cop* highlights the latent tensions between these two factions. In the film, both sides are given ample opportunity to delve into clichés toward each other. Yet the plot goes beyond the level of prejudice to, as Völkl proposes, not only construct, but also deconstruct linguistic, political, and sociocultural boundaries between the two groups, suggesting that cooperation may be the best option, after all. Whereas Völkl focuses on inner-Canadian relations, Evelyn P. Mayer investigates the relationship between Canada and its southern neighbor, the United States. As Mayer's »The Significance of the United States and the Canada–U.S. Border for Canadian National Identity Construction« suggests, Canadian identity is always constructed vis-à-vis the United States. Taking instances from Canadian literature as well as popular culture, Mayer traces Canadian identity as a contested, uneven terrain, in which Canadians increasingly seem to feel the need to insist on a self forged independently from their U.S.-American neighbor.

The final part of this collection tackles the space of the marginal in literature and popular culture. Diana Benea's chapter discusses the haunted presence of Native Americans in American society as depicted in Thomas Pynchon's novel *Mason & Dixon*. Benea suggests that by positing Native Americans as ghosts within the story, Pynchon critically assesses America and American history as spaces characterized by a ›return of the repressed‹, in this case the repressed trauma of Native American genocide. Madalina Prodan's »Searching for a Mixed-Blood Identity in Sherman Alexie's *Flight*« equally deals with the issue of Native American historiography (or lack thereof) within American literature. Her chosen text, Sherman Alexie's novel *Flight*, focuses on the coming-of-age of a young mixed-blood. Deconstructing conventional notions of Indianness, Alexie, according to Prodan, presents the hybrid identity of the mixed-blood as a potential mediating force between the dominant white and the marginalized Native American cultures. In the volume's final essay, Judith Kohlenberger traces science's development from a marginal space of freaky nerds toward a money-making industry. Through various popular broadcasting formats appealing to the masses, science has not only become more accessible to a wider audience, it has also become associated with coolness rather than ›nerdsville‹. As Kohlenberger argues, science and popular culture have, in fact, become mutually constituent spaces.

There is definitely much more to be said about space (and time) in America, and many more directions and angles could be chosen for such a collection. But while the space of thought and critical inquiry ideally remains an open-ended field, the space of printing and publication, unfortunately, is not. One thing seems rather certain: As a point of departure for scholarly analysis and discussion, space is here to stay, and our little book seeks to do its part in furthering that discussion and in keeping it open to various perspectives.

Notes

1 | While I am specifically using a quote by a member of the conservative faction, I, of course, in no way want to suggest that Amy Kremer or representatives of the Tea Party or the Republican Party have been the only ones capitalizing on this idea in the course of the campaigns for the 2012 elections. Images linking contemporary America to the so-called ›Founding Fathers‹, thereby underscoring America's preeminence as ›world leader‹, were utilized by both Democrats and Republicans alike.
2 | I am using the present tense here simply for clarity and succinctness. Whether we actually are still located in postmodernity is, of course, up for debate.

References

Anderson, Benedict (1983/1991): *Imagined Communities. Reflections on the Origins and Spread of Nationalism*, Revised Edition. London: Verso Books.
Anzaldúa, Gloria (1987/1999): *Borderlands/La Frontera: The New Mestiza*, 2nd Edition. San Francisco: Aunt Lute Books.
Atwood, Margaret (1972/2012): *Survival: A Thematic Guide to Canadian Literature*. Toronto: House of Anansi Press.
Baudrillard, Jean (1986/1999): *America* (trans. Chris Turner). New York: Verso Books.
Benesch, Klaus (2005): »Concepts of Space in American Culture: An Introduction«, in Klaus Benesch/Kerstin Schmidt (eds.), *Space in America: Theory–History–Culture*. Amsterdam: Rodopi, pp. 11–21.
CNN (2012): »Starting Point with Soledad O'Brien«, *CNN* [online], 4 September, http://transcripts.cnn.com/TRANSCRIPTS/1209/04/sp.02.html. 9 September 2012.
Kennedy, John F. (1961/2012): »Massachusetts General Cout, January 9, 1961« [Address of President-Elect John F. Kennedy Delivered to a Joint Convention of the General Court of the Commonwealth of Massachusetts], *John F. Kennedy: Presidential Library and Museum* [online], January 9, http://www.jfklibrary.org/Asset-Viewer/OYhUZE2Q0o-ogdV7ok9ooA.aspx. 9 September 2012.
Lefebvre, Henri (1974/2000): *The Production of Space* (trans. Donald Nicholson-Smith). Malden: Blackwell Publishers.
Morley, David (1996): »EurAm, Modernity, Reason and Alterity; or, Postmodernism, the Highest Stage of Cultural Imperialism?« in David Morley/Kuan-Hsing Chen (eds.), *Stuart Hall: Critical Dialogues in Cultural Studies*. London: Routledge, pp. 392–408.
Said, Edward (1993): *Culture and Imperialism*. New York: Vintage Books.

Soja, Edward (1996): *Thirdspace: Journeys to Los Angeles and Other Real and Imagined Places*. Malden: Blackwell Publishers.

Spanos, William (2003): »A Rumor of War: 9/11 and the Forgetting of the Vietnam War«, *boundary 2* 30 (2), pp. 29–66.

Winthrop, John (1630/1994): »A Model of Christian Charity«, in Paul Lauter (ed.), *The Heath Anthology of American Literature*, Volume One, 2nd Edition. Lexington: DC Heath and Company, pp. 226–234.

Zinn, Howard (1980/2005): *A People's History of the United States: 1492–Present*. New York: Harper Perennial.

Constructing America from Afar

Performing America Abroad
No Name City and the Haunted Spaces
of Transnational America

LEOPOLD LIPPERT

Over the last two decades, much theorizing about the scope and methods of American Studies, routinely acknowledged by its practitioners as a vibrant and interdisciplinary field of inquiry, has been set apart by a sense of urgency. The increasingly uninterrupted global flows of capital, goods, ideas, images, technologies, and people—a process that is commonly referred to as ›globalization‹—have compelled American Studies scholars to adjust their lines of research. Already in 1996, Jane C. Desmond and Virginia R. Domínguez demanded a »concerted effort throughout the American Studies scholarly community to embrace actively a paradigm of critical internationalism as we move into the next century« (1996: 475). The following years saw much thinking about what *critical internationalism* may mean, and a rather extensive exploration of the many tensions between the local and the global, the national and the international, as well as between modernity and its aftermath. At the same time, the study of border crossings, modes of exchange and dialogue, and spaces of contact have enabled a more thorough understanding of the social and historical techniques that produce notions of locality and globality in the first place.[1]

In her influential 2004 presidential address to the American Studies Association (ASA), Shelley Fisher Fishkin could already turn to a vast body of scholarship concerned with »looking beyond the nation's borders, and understanding how the nation is seen from vantage points beyond its borders« (2005: 20). Acknowledging the productive collaborative enterprise that has brought together scholars from diverse regional and institutional settings around the globe, Fishkin boldly announces the »transnational turn in American Studies« (2005: 17) and calls on her colleagues to »see the inside and outside, domestic and foreign, national and international, as interpenetrating« (2005: 21). Fishkin's pronounced insistence on the importance of the *transnational turn* has stimulated much scholarship in recent years, and the 2009 inauguration of the high-profile *Journal of Transnational American Studies* stands as a testament to an indispensable line of research.

While it is virtually impossible to review every single piece of scholarship concerned with the transnational in American Studies, I want to suggest that research so far has been going mainly into three directions. First, there is the study of *Americanization*, that is, the (mostly) unidirectional spread of U.S. culture around the globe. An attempt to compensate for what Amy Kaplan has rightly identified as »the absence of empire from the study of American culture« (1993: 11), the Americanization paradigm foregrounds the enormous cultural and political power of the United States on a global scale.² Second, much work has adopted what Günter H. Lenz calls a »dialogic notion of cultural critique« (2002: 474). Dismissing the idea of simple Americanization, Lenz argues that »[i]t is not sufficient in a time of globalization to trace the cultural impact of the United States on other nations all over the world in order to reintroduce the history of American imperialism into the debates about postcolonialism« (2002: 472). This second approach, then, highlights the fundamental *dialogics* of cultural transformation and introduces a more balanced view of the role of Americanness³ in an age of globalization. Built on the ideas of encounter and exchange, the dialogic points to the agency not only of the United States, but of a larger number of nation-states and national subjectivities. Third, the concept of *hybridism* accentuates the curious blends, mixtures, and intercultures that emerge from these transnational dialogues. Relying on the postcolonial concept of the ›Third Space‹, this »split-space of enunciation« (Bhabha 1994/2004: 56) that impedes any smooth closure of signification to begin with, American Studies scholars pursuing this third approach focus on metaphorical and literal borderlands and crossroads, and insist on Homi Bhabha's proclamation that it must always be »the ›inter‹—the cutting edge of translation and negotiation, the *inbetween* space—that carries the burden of the meaning of culture« (1994/2004: 56).⁴

No Name City and Americanness in Austria

With these roughly outlined theoretical approximations in mind, Florian Flicker's 2006 documentary *No Name City: Die Freiheit liegt südlich von Wien* [›Freedom Lies South of Vienna‹] can be considered a cultural product particularly useful for the examination of transnational Americanness. The ninety-minute film chronicles one summer at No Name City, a Western-themed entertainment park located near Wöllersdorf, thirty kilometers south of Vienna. Flicker's documentary starts out with a rehearsal: The town staff, already dressed up as

Cowboys, Indians[5], or Sheriffs, prepares for an elaborate Wild West show that features a bank robbery, street shooting, an Indian horse show, and the ambushing of the miniature train that chugs through No Name City. After this opening sequence, complete with an impromptu performance of the Austrian *Schlager* singers Waterloo and Robinson, the film continues to detail the more mundane work routines of the town's employees: Flicker sits in the car with the ›garbage man‹ Sunshine Kid, associates with the gardener, chats with the owner of the local bar, visits the proud owner of a Western clothing and accessories store, goes over horsewoman Michaela's busy schedule, and talks about the economic situation of the *Westernstadt* with Armin Groß, the manager of No Name City. During Flicker's meanderings through the town, the audience sees not only railway tracks, settler tents, cowboys, and Western saloons, but also the United States flag, the Texan flag, and the Confederate flag: the symbolic visual language of the nineteenth-century U.S. nation-building project is omnipresent.

While the film starts out as a relatively straightforward documentary about an especially remarkable manifestation of Americanness in Austria, *No Name City* (and its filmmaker, who is often present among his protagonists) soon becomes entangled in a severe confrontation between management and employees of the town. In a precarious financial situation, the new supervisor Armin is dissatisfied with the staff's work performance and reluctance to his improvement plans. On several occasions, he threatens to lay off employees and »do it all by [him]self«.[6] The town staff, on the other hand, is disconcerted with Armin's harsh managerial style—led by the vocal Michaela, they demand participation rights and direct acknowledgment of their hard work. As the conflict steers toward escalation, the Austrian singer and entertainer Hans Kreuzmayr, better known to his audiences as Waterloo, pays a visit to the *Westernstadt*.[7] Waterloo, a former Eurovision Song Contest participant who has made his (late) career performing Indianness on various stages, is regularly hired to *play Indian* in No Name City.[8] When he appears in these climactic scenes of the film, like a veritable *deus ex machina*, Waterloo is already dressed up as German novelist Karl May's famous Indian character Winnetou and rides into the town on his horse. Adopting the conciliatory personality of the fictional Winnetou (who is, in Karl May's version, the amicable Chief of the Apaches), Waterloo promises to make peace between the enemy factions: In what can be considered an appeasement effort, he talks to staff leader Michaela, reasons with manager Armin, and even boldly announces to use his influence with the Lower Austrian governor Erwin Pröll.

In an extended scene, finally, Waterloo tries to call Bernhard Negedly, the owner of No Name City, in order to ask him for support. Kreuzmayr takes his cell phone out of his Indian buckskin leather clothes and, fully aware that his performance is being filmed, makes the call (see *Illustration 1*). Unfortunately, he doesn't get any further than to Negedly's secretary, who declares that Neg-

Illustration 1a and 1b: Waterloo trying to reach Bernhard Negedly, the owner of No Name City, via cell phone

Source: *No Name City* DVD. *No Name City* © Hoanzl, 2006.

edly is out of the country and cannot be reached. As it turns out, the failure to contact Negedly is already foreshadowing the larger failure of Waterloo's project: Although he leaves the town in the final shot of *No Name City* with a repeated invocation of his optimistic mantra, »together we are strong,« the subsequent inserts inform the audience that Kreuzmayr's attempts proved futile. Several of No Name City's employees left the town soon after filming was over, and the conflict was not so much resolved as cut short by the staff's disillusioned departure.

Transnational Americanness seems to be everywhere in this film about a small *Westernstadt* south of Vienna. One could certainly think about *No Name City* along the lines of Americanization, and investigate the spread of U.S. national mythology, historical practices, and the symbolism of the Wild West into Austrian popular performance culture. One could, of course, further complicate the picture by focusing on the construction, popularization, and standardization of these very Wild West narratives through the productions of *Buffalo Bill's Wild West*, a highly successful late nineteenth-century entertainment business traveling the globe.⁹ One could certainly also frame the documentary in a more dialogic way, and foreground the influence of what Richard H. Cracroft has called »the American West of Karl May« (1967: 249), that is, the fictional engagement of a German author with Americanness, on *Westernstadt* performance practices. One could also highlight transatlantic encounters and dialogues by focusing on »Indianthusiasm« (Lutz 2002: 167) in Austria, a fervent passion for all things Indian that has generated extensive exchange networks between hobby groups and performers from Europe and Natives from the United States. One could also attempt to untangle the shared fabrication of Indianness by both Native Americans and the white U.S. mainstream and explore how these hybrid cultural constructions are played out on a transnational stage.[10] By closely reading *No Name City* and following Fishkin's recommendation to *look beyond the nation's borders*, then, one could certainly gather new insights into the global production of Americanness.

Disciplinary Challenges

I would caution, however, against such narrow understandings of the transnational, which are situated along axes of difference between nations and national cultures, and which concentrate only on routes of exchange and spaces for dialogue. I would also caution against a comparatively uncritical embrace of hybridism and a scholarly showcasing of its transformative potential. As Gudrun Rath (2010) has pointed out, the idea of ›the third‹, due to its dependence on a preceding dichotomy, always runs the risk of becoming an essentialist, more or less stable category itself.[11] Although these established approaches all have their benefits and certainly prove worthwhile for a good number of research perspectives, they preclude a more sustained consideration of the larger epistemological and ontological changes brought about by contemporary processes of globalization. These changes, Arjun Appadurai insists, have to do with the gradual disappearance of the national as one contemporary epistemic framework. »As units in a complex interactive system«, Appadurai claims, »[nation-states] are not very likely to be the long-term arbiters of the relationship between globality and modernity. That is why [...] I imply that modernity is at large« (1996: 19). George Lipsitz strikes a similar chord: while he maintains that national narratives do not simply disappear in a *modernity at large*, in a postindustrial globalized world, he still reasons that much of the »institutional American studies« depends on »connections between culture and place that no longer may be operative« (2001: 27). For Lipsitz, the challenge to American Studies is to come to terms with the realization that new spatial relations also produce new ways of knowing and being. In an increasingly transnational realm, he claims, scholarship has to acknowledge the emergence of »new social subjects who inevitably create new epistemologies and new ontologies« (2001: 8).[12]

In this sense, I would argue that the transnational has much more to do with changes and shifts in knowledge relations than with the study of encounters and exchanges between nations. The attention given to the transnational turn, I propose, could also be seized as an opportunity to explore *post*national knowledges rather than *inter*national contact, and to think about cultural processes *beyond* the national as a framework for the production of subjectivity. As a movement away from the nation-state, the transnational should be considered an approach rather than an object of study. It is a conceptual vehicle that is always in transit, an evasive maneuver that informs and transforms every routine American Studies scholars conventionally perform to study what Lauren Berlant has famously called the ›National Symbolic‹.[13] Instead of conceptualizing the transnational as a distinct subsection of the field, it should be considered a transdisciplinary exercise, an unruly performance intervening in the disciplinary setup of American Studies as an institutionalized academic project.

It would be misleading, then, to subscribe to Fishkin's concession that despite the significance of the transnational turn, there is still much to do for a distinctly ›national‹, institutionalized American Studies. When Fishkin hedges, »I don't want my remarks here to suggest that everyone needs to do transnational work« (2005: 22), she unwittingly deflects from the larger issue, namely that in a globalized, postmodern world, there is little other work left. As Michael Hardt and Antonio Negri so rightly hold, »[i]t is false, in any case, to claim that we can (re)establish local identities that are in some sense *outside* and protected against [...] global flows« (2001: 45). Philip Deloria's 2008 back-to-business, or rather, back-to-›Broadway and Main‹, address to the ASA is puzzling in a comparable way. He relegates the »transnational and global turns« to one of »four recent thematic interests«—firmly placed within the field—along with »race and ethnicity studies, [...] community scholarship, and [...] civic engagement« (2009: 2). While Deloria treats the transnational as yet another fashionable area of expertise among many and urges Americanists to circle back to ›Broadway and Main‹, I want to warn against the questionable sense of security that comes with disciplinary self-confidence. It seems to me that the challenge posed to American Studies scholars is not so much to return to Broadway and Main, but to come to terms with the global, and take seriously the transnational, not as an area of study, but as a practice—the radical rethinking of the institutional and methodological premises of the field.

It might have become clear already that the debates around transnationalism and American Studies strike at the heart of the discipline and its institutional apparatuses. Although often lauded as *inter*disciplinary and welcoming to almost every kind of scholarship, American Studies has always depended on a national reference system localized with some connection to the United States. Even the search for Americanness in the wider world has always implied the transportation and adaptation of national vernaculars to global settings. American Studies, historically speaking, represents a distinctly modern project: a securely institutionalized endeavor to enable the highly contested yet thoroughly disciplined production of national subjectivities and American bodies. It is no coincidence then that Lucy Maddox's attempt to ›locate‹ American Studies in a widely-read collection of essays »chosen to represent the development and growth of American studies as an academic enterprise devoted to the interdisciplinary study of American history and culture« (1999: vii) is subtitled *The Evolution of a Discipline*. It is also no coincidence that Judith Halberstam, in her response to Deloria's already mentioned presidential address, conjures the spirit of Michel Foucault in order to reveal the (unacknowledged) disciplinary and disciplining assumptions undergirding his argument. Halberstam urges her colleagues at the ASA to »take this opportunity to rethink the project of disciplinarity altogether« (2009: 34) and imagines future scholarship to »detour around the crossroads or interdiscipline, [to] deterritorialize American studies; [to] refuse its boundaries and [to] break with the gatekeepers of the disciplines« (2009: 36). Halberstam's intervention is crucial, as it points to the often overlooked importance of institutional and disciplinary legacies for the conception and reconception of theoretical frameworks and methodological expectations.

Already ten years earlier, Janice Radway, in her 1998 presidential address to the ASA, »What's in a Name?« tried to undertake one such detour, however only to be circling back to the confines of a (reanimated) disciplinarity. In this address, Radway attends to the sudden disappearance of America in a transnational age and ponders possible names for both American Studies and the ASA at a time when »the notion of a bounded national territory and a concomitant national identity deriving isomorphically from it are called into question« (1999/2002: 59). Like Appadurai, Lipsitz, and others, Radway realizes that processes of globalization and post-industrial migration and mediation pose a fundamental problem to a nation- or area-based discipline like American Studies. Attempting to resolve the issue, she proposes three different names (standing in for methodological perspectives) for American Studies as a field. Suggesting the reductive »United States studies« (1999/2002: 60), the hemispheric »Inter-American Studies« (1999/2002: 62), or the deterritorialized »Intercultural Studies« (1999/2002: 64), Radway indicates a number of directions into which the discipline could be headed. In the end, however, she dismisses, without much argumentation, all three approaches and instead opts for traditional, though reconfigured, American Studies methodology. Conceiving of a project that makes it »possible to honor the past and build on the successes of the field […] even while mounting a responsible but vigorous critique of past myopias and earlier paradigms« (1999/2002: 65), Radway closes with a clichéd pledge of allegiance to the discipline. »I do not think this field or this association need to fear change,« she states. »Together they have fostered it in the past and embraced its effects; they can do so again« (1999/2002: 65).

Although it is, of course, difficult to speculate about her motivations, Radway's refusal is significant, as it points to the ongoing importance of the national as a modern, disciplinary category in American Studies scholarship. Radway embraces the established institutional landscape at the expense of a more sustained search for alternatives to what she herself has condemned earlier as the all too easy adoption of an »ontologically prior American identity« (1999/2002: 58). But what would a post-disciplinary, transnational *academic enterprise* look like? What would it mean to evade disciplinary knowledge production and follow Judith Halberstam's not so tongue-in-cheek advice taken from the movie *Madagascar* to »get lost, stay lost« (2009: 37)? What rethinking would it involve to understand the transnational as a theoretical move, as a scholarly performance rather than a field of study? My suggestion would be to think about the debates around transnational American Studies in the context of larger epistemic changes in contemporary global culture. As Jon McKenzie has so persuasively argued, there is currently happening a shift from *discipline* to *performance* as primary modality for the production of knowledge. Such a shift gradually establishes the predominance of performance, in Diana Taylor's words, as »an episteme, a way of knowing, not simply an object of analysis« (2003: xvi). As a consequence, I will proceed to examine the transnational in American Studies as part of these epistemological (and ontological) changes and approach it simply for what it is: a performance.

FROM DISCIPLINE TO PERFORMANCE: THE GHOST OF AMERICA

Jon McKenzie opens his book *Perform or Else: From Discipline to Performance* with a startling observation: »Today, as we navigate the crack of millennia,« he remarks, »work, play, sex, and even resistance—it's all performance to us« (2001: 3). Pondering the ubiquity of the concept of *performance* in a number of fields seemingly unrelated to his »native« Theater and Performance Studies (not to mention popular discourse), McKenzie embarks on a daunting project: »to rehearse a general theory of performance« (2001: 4). A general theory, for McKenzie, would make sense of performance not only as a descriptive term associated with cultural performance events on a spectrum ranging from ritual to theater, but would encompass a much broader terrain: the paradigm is being used, after all, in connection with organizational and technological performances as well. In order to illustrate briefly and somewhat humorously the curious combination of a number of seemingly separate performance practices, McKenzie turns to a cover page of *Forbes* magazine, the very cover that inspired the title of his book. The cover photograph shows an action well-known to scholars of theater, as it is a quotation from the vaudeville stage: »a cane hook wrapped menacingly around the neck of a white-skinned, gray-haired businessman« (2001: 4–5). The performance the cane indicates, however, is not only cultural, but part of yet another performance paradigm, revealed by the magazine's headline: »Perform—or else: Annual Report on American Industry« (2001: 4). Pointing to managerial or organizational performance, McKenzie claims that »the *Forbes* challenge and its hold upon throats around the world« sure is dramatic, but not in the sense of a vaudeville routine—it reads as follows: »Perform—or else: be fired, redeployed, institutionally marginalized« (2001: 7). But there is still a third performance principle to be discerned on the *Forbes* cover page: it is the bar code with machine-readable information about the magazine's price. »Bar code«, McKenzie argues, »is a script of technological performance, a performance embodied in such items as high performance sports cars, stereos, and missile systems« (2001: 10). Obviously, this performance of technology is crucial for the others to work as well. Performance, then, seems to permeate the social and cultural landscape on a larger number of terrains than one would expect and informs the production of cultural knowledge in a variety of intriguingly interrelated ways.

In *Perform or else*, McKenzie speculates that the odd recurrence of performance in seemingly distinct contexts and disciplines points to a larger epistemic shift that has been underway since the end of World War II. He provocatively asks, »What if the diversification and proliferation of researchers, projects, and fields over the past fifty years signal not only a quantitative leap in research initiative, but also a qualitative mutation in what we call knowledge, the becoming-performative of knowledge itself?« (2001: 13–14). McKenzie's general theory of performance proposes nothing less than a modification of thinking about knowledge. The ongoing transition from modernity to something beyond, he claims, has been accompanied by a transition in epistemology and power as well. While Michel Foucault has shown, most comprehensively in his *Discipline and Punish* (1975), that the disciplinary regime and its specific technologies of power have been at the center of the *modern* Western knowledge project, McKenzie claims that this is no longer the case for a contemporary, *post*modern world. Instead, he suggests that »*performance will be to the twentieth and twenty-first centuries what discipline was to the eighteenth and nineteenth, that is, an onto-historical formation of power and knowledge*« (2001: 18). An epistemic shift from discipline to performance, accordingly, would necessitate a gradual desertion of disciplinary modes of power and systemic knowledge, so treasured by the institutions of modernity. In the age of postmodernity, bodies and subjectivities are being produced increasingly by the knowledges and technologies of performance.

But what is the epistemology of performance? What does performance signify other than the obvious—the embodiment of behavior? And what would a study of the transnational performance of ›America‹ encompass? A number of scholars have attempted a definition: Marvin Carlson, in an extensive overview of writings on performance, points to a crucial characteristic: »[P]erformance,« he claims, »is associated not just with doing but also with re-doing« (1996/2004: ix). As this observation makes clear, performance must be conceived of as immanently processual. The concept is associated not only with embodied behaviors but with the »embodiment of the tension between a given form or content from the past and the inevitable adjustments of an ever-changing present« (1996/2004: ix). To study performance involves the study of the *tensions* between different versions of embodiment—the examination not only of ›doing‹, but of why, when, where, and in what ways things are ›re-done‹, and re-done differently. Richard Schechner develops a related idea of performance: In his influential *Between Theater and Anthropology*, he describes it as »restored behavior« (1985: 35). For Schechner, »[p]erformance means: never for the first time. It means: for the second to the *n*th time« (1985: 37). Performance, thus, serves as a powerful echo of an elusive source that can never be identified exactly; it is a constant and complex restoration of an irretrievable »first time«, a lost original. As Joseph Roach points out, performance always involves »surrogation« (1996: 2), that is, the continuous and always already thwarted effort to re-do what has never been done for the first time. »Performance, in other words,« Roach insists, »stands in for an elusive entity that it is not but that it must vainly aspire both to embody and replace« (1996: 3).

As *n*th-time embodiment, performance must be understood as a continuous reenactment that is never a literal repetition of existing social structures or cultural constellations; instead, performances constantly actualize potentialities. The epistemology of performance is not concerned with policing and transgressing the boundaries of pre-established and institutionalized disciplinary knowledge, but with the inherently self-referential feedback loops created by performance's restorations from the *second to the* n*th time*.[14] It is only concerned with the echoes of America surfacing in processes of globalization, not with the disciplinary matrices of an American Studies practice that has so casually connected place, origin, and culture. The epistemology of performance, then, allows for the study of the transnational as a processual, open-ended phenomenon that no longer depends on a set of original American characteristics stipulated by the technologies of the discipline.

But performance is not only an endless, inconclusive, self-referential process: it also points to an ontological failure. As Peggy Phelan explains, »[p]erformance in a strict ontological sense is nonreproductive. [...] Performance clogs the smooth machinery of reproductive representation« (1993: 148). For Phelan, performance is of a more shadowy ontology associated with a crisis of knowledge, as it is always contingent on its failure to reappear exactly the same: »Performance's being [...] becomes itself through disappearance« (1993: 146), she thus describes the ontology of performance. This line of thinking bears a striking resemblance to Jacques Derrida's reflections on the *Specters of Marx*, developed from a plenary address delivered at the 1991 »Whither Marxism?« conference at the University of California, Riverside. Although Derrida pursues a different political project that involves a specific politics of mourning as reparative gesture toward history's forgotten subjects, I am interested here particularly in his concept of *hauntology*. Derrida introduces the *specter* as »this non-present present, this being-there of an absent or departed one« (1993/2006: 5). The specter, he claims, »no longer belongs to knowledge. At least no longer to that which one thinks one knows by the name of knowledge« (1993/2006: 5). Like performance, the specter is also predicated on its disappearance, on its present absence, and the crisis of knowledge that accompanies it; like performance, the specter proposes a different ontology. It is precisely the specter, this strangely *non-present present*, what figures in the restorations of performance. The untimely (dis)appearances of ghosts, of *first times* never exactly determinable, then, are haunting the performance processes that create cultural knowledge. For haunting, says Derrida, »is historical, to be sure, but it is not *dated*, it is never docilely given a date in the chain of presents« (1993/2006: 3).

The question, or rather, the challenge posed by the ontology of performance is thus: Whither *America*? And whither *American Studies*? For an examination of how America is being performed abroad, how Americanness goes transnational, it might not be appropriate to discard at all the nation-based idea of America; rather, I find it useful to speak of the ghost of America, a strangely historical formation reminiscent of older, disciplinary thought systems that is haunting the contemporary cultural landscape. America, in this sense, refers to the performative *being-there of a departed one*, an old friend, or foe, that is no longer entirely graspable, of a different ontological order. According to Avery F. Gordon, ghosts are experienced »not as cold knowledge, but as a transformative recognition« (1997/2008: 8). Through performance, then, the ghost of America manifests as its inherent transformation: Its (dis)appearances are contingent on its performative restorations and surrogations, the always already frustrated re-doings of a lost *first time*, not on seemingly *original* national categories. Like a faint echo, the ghost of America is haunting the channels of global cultural production, presenting a challenge not to hold on the paradigms of American Studies as a discipline, but to constantly reenact, rephrase, and rework the power/knowledge matrix assembled around that elusive entity called Americanness. For American Studies scholarship, the task is to decode these ghostly messages, and to respond to transnational performances *beyond* a broad range of disciplines. The challenge, then, is not to come up with conclusions, or with a secure idea of what Americanness may mean, but a challenge to study transnational performances for the second time, or for the *n*th time, never for the first.

NO NAME CITY AS A HAUNTED *WESTERNSTADT*

Thinking about *No Name City* as an example of how transnational Americanness is *performed* abroad, I would suggest to pay attention to the several failures the film chronicles. In fact, the documentary seems to be less about Americanness in Austria than about the unattainability of some preconceived notion of Americanness and the disappearance of America in a global performance conundrum. What is at stake in *No Name City* is neither genuine transatlantic dialogue nor the authentic restoration of America in an Austrian *Westernstadt*; what is at stake, rather, is an engagement with the ghost of America—a haunting presence that is constantly restored, but that always fails to live up to an *original* in the first place. With that in mind, I find it important to recall McKenzie's example about the threefold performance of *Forbes* magazine. For the transnational performance of *No Name City* is not merely cultural, but is composed of a larger web of entangled, interdependent performances that are managerial, technological, *as well as* cultural. Thinking about No Name City as a haunted *Westernstadt* necessitates an understanding of the interactions of the various layers of performance, their restorations and their disappearances, all of which jointly conjure the ghost of America in a small town south of Vienna.

Remember that Flicker's documentary starts out with a rehearsal. Three male employees of No Name City, already dressed up as cowboys and sheriff with guns, hats and heavy boots, are sitting in one of the town's bars, thinking over their lines and movements for the following Western show. The planned choreography is elaborate: the actors envisage a dramatic performative recreation of a bank robbery complete with a shooting in the town's main street. When it comes to the actual performance in front of a small audience, however, the shooting scene is harshly interrupted. A woman who also works for the town crosses the street in a horse wagon full of supplies. As the gunfire scares her horses, she breaks the frame of the Western performance and addresses one actor, »Just let me pass, please.«[15] Of course, the supply drive, necessary for the city's organizational performance, ruins the shooting scene, which is interrupted and resumed only after a short break. A similar failure affects the small train that is supposed to ride visitors through the town as part of the Western show. At one point in the film, it breaks down, forcing the many children who boarded already to skip this particular performance of America. While a repair crew attempts to get the train going again—without much success—, the audience witnesses the technological failure of a vehicle so central to nineteenth-century U.S. westward expansion. A third performance of Americanness in the transnational realm proves unsuccessful as well: When Waterloo and his former singing partner Robinson reunite to perform their bilingual song »Meine Kleine Welt/My Little World« in front of Flicker's camera, their rendering is far from perfect. Although they are filmed sitting on a bench perfectly positioned against the romantic backdrop of the *Westernstadt*, although Waterloo wears his Indian clothes and Robinson has donned a cowboy outfit, their performance seems doubtful: the singers mix up their lines, their thirds are out of tune more often than not, and they are unsure at which points to switch from German to English. To complete the disorderly picture, the dog which has been sleeping under the bench gets up at one point and walks through the frame. The spectacular attempts of the *Westernstadt* performers to recreate an American mythical landscape are performances in an epistemic sense: interrupted by the mundane transport of supplies through the town, by the breakdown of a train, and by the disturbing presence of a straying dog, they are effecting not so much a recreation of original Americanness as its constant relegation to a different ontological order—to that of *disappearance*. Performing America abroad, then, involves its disintegration; it is made manifest only as ghostly presence, always already absent.

In addition, the larger setup of the film points to the performative disappearance of America into the transnational workings of a more globalized capitalism. Although *No Name City* begins as a documentary about the exploration of American history in Austria through cultural performance, it soon becomes clear that there is a different performance which is much more central to the *Westernstadt* and its inhabitants. As Flicker laments in a statement recorded during the time of filming (featured as bonus material on the DVD of *No Name City*), »We have been here for two weeks now and are a bit puzzled because of the disagreement among the city's inhabitants. And because we don't know how we can get the people back together in order to shoot our film.«[16] As it turns out, Flicker decided to shift his focus and make the conflicted organizational and managerial performances the central theme of his film. *No Name City* makes clear that life in the *Westernstadt* is not so much about an ›authentic‹ recreation of Americanness or about ›sincere‹ transatlantic dialogue, but about the fundamental ability to *perform* at all, that is, to keep the business operation No Name City alive administratively and financially. This becomes particularly evident in the film's climactic moments, in which Waterloo's thwarted attempt at performing the American myth of Indianness illustrates a multilayered transnational performance haunted by the ghost of America. Dressed up as the peace-loving Karl May character Winnetou, Waterloo adopts the conciliatory personality of the Apache chief and calls the owner of No Name City, Bernhard Negedly, for help. While this scene represents the culmination of an everyday conflict in global, postindustrial capitalism, and while it depicts primarily how the technological performance of a cell phone is used to solve a problem concerning the organizational performance of the *Westernstadt*, it is still strangely haunted by the ghost of America. During the entire sequence featuring Waterloo, his Indian persona is never entirely dropped, as he uses its ghostly presence as an argumentative rationale for his ambitions as a peacemaker. Waterloo's repeatedly expressed managerial motto, »together we are strong« is haunted by the non-present presence of Americanness channeled through Karl May's famous character Winnetou. In the end, however, the audience witnesses a performance breakdown both of the Indian effort to make peace and of the managerial effort to survive. As the inserts that conclude the documentary indicate, a considerable number of the *Westernstadt* employees left disillusioned after the summer; and a quick search on *Wikipedia*, yet another one of these performative, inherently processual knowledge production machines, reveals that No Name City itself has been defunct since 2008 (Wikipedia 2011).

Understanding *No Name City* as performance, I thus take seriously the transnational as a mode of knowledge, as a practice, not as a field of research always already disciplined by the methodological imperatives of a nation-based American Studies. Like all performance, transnational Americanness »resists the sort of definitions, boundaries, and limits so useful to traditional academic writing and academic structures« (Carlson 1996/2004: 206). Never attempting to access an ostensible national origin, transnational Americanness travels the cultural circuits *beyond* the nation, participating in the performative restoration of an America that can never be recreated exactly. As a consequence, the globalized culture presented in *No Name City* is haunted by the strange presence of the ghost of America, a ghost which echoes the faint howls of lost origins, national boundaries, and disciplinary epistemologies. Speaking to this ghost, then, involves the particular challenge posed by the power/knowledge matrices of a postmodern, transnational world: it means to enter the epistemology of performance.

Notes

1 | Work on this chapter has been made possible by a Ph.D. fellowship (›Forschungsstipendium 2011‹) awarded by the University of Vienna.

2 | A number of works concerned with questions of empire and Americanization can be found in Amy Kaplan and Donald E. Pease's collection *Cultures of United States Imperialism* (1993), for example Mary Yoko Brannen's »›Bwana Mickey‹: Constructing Cultural Consumption at Tokyo Disneyland«. With regard to Austria, I would single out Reinhold Wagnleitner's *Coca-Colonization and the Cold War* (1991/1994), although itself critical of the term ›Americanization‹, as an important and meticulously researched study.

3 | Whenever I use ›America‹ or ›Americanness‹ in this chapter, I am referring to a highly contested cultural construction rather than to an inherent set of national characteristics. Although this is usually indicated by quotation marks (or »scare quotes« [2010: 5], as Bryce Traister complains), I will refrain from this practice here so as not to unnecessarily complicate the reading process.

4 | Apart from Bhabha's *The Location of Culture* (1994/2004), American Studies scholars frequently rely on Gloria Anzaldúa's *Borderlands/La Frontera* (1987/1999) or Mary Louise Pratt's »Arts of the Contact Zone« (1991) as theoretical background. Important research combining the dialogical approach with discourses of hybridism can be found, for example, in Paul Giles' *Virtual Americas* (2002), Jaap Kooijman's *Fabricating the Absolute Fake* (2008), or in Astrid Fellner's »Crossing Borders, Shifting Paradigms« (2008).

5 | Whenever I use ›Indians‹ or ›Indianness‹ in this chapter, I refer to the socially mediated cultural representation of Native American characteristics and people. The ascription ›Indian‹, as well as, of course, the similarly problematic designation ›Native American‹ itself, must be understood always as a legacy of violent and unjust colonial power dynamics. For reasons of readability, I will refrain from using quotation marks to indicate this.

6 | »[Wenn nicht], mach ich alles selber.« All translations from German are mine.

7 | While Waterloo is not the central figure in this argument, I have written in more detail about Kreuzmayr and the politics of his ›Indian‹ persona in my »Playing Indian in Austria: Waterloo and the Transnational Performance of America« (2012).

8 | I borrow the phrase ›playing Indian‹ from Philip J. Deloria's book of the same name. In this detailed history, he argues that ›Indianness‹ in the United States has always been an issue of playing, of performing particular and often contradictory fantasies. »The donning of Indian clothes«, Deloria claims, »moved ideas from brains to bodies, from the realm of abstraction to the physical world of concrete experience. There, identity was not so much imagined as it was performed« (1998: 184).

9 | For two excellent (and complementary) books on the representational politics and historical significance of *Buffalo Bill's Wild West* shows, see Joy S. Kasson's *Buffalo Bill's Wild West* (2000) and L. G. Moses' *Wild West Shows and the Images of American Indians, 1883-1933* (1996).

10 | For insightful accounts of the complicated constructions of Indianness in the United States, see especially Robert F. Berkhofer, Jr.'s *The White Man's Indian* (1979), S. Elizabeth Bird's edited collection *Dressing in Feathers* (1996), and two books by Philip J. Deloria, *Playing Indian* (1998) and *Indians in Unexpected Places* (2004).

11 | Rath's essay »›Hybridität‹ und ›Dritter Raum‹: *Displacements* postkolonialer Modelle« (2010) is part of a collection that critically assesses the uses and misuses of the figure of the third for the study of cultural processes.

12 | Although not explicitly addressed in this context, Lipsitz' argument is built on the debates about postmodernity and space led primarily in the 1990s. See, for example, Edward W. Soja's *Postmodern Geographies* (1989), David Harvey's *The Condition of Postmodernity* (1990), Roland Robertson's *Globalization* (1992), and Lipsitz' own *Dangerous Crossroads* (1994). Judith Halberstam has taken up the issue from a queer perspective in her *In a Queer Time and Place* (2005).

13 | In her study of Hawthorne and American national fantasy, Berlant refers to the ›National Symbolic‹ as »the order of discursive practices whose reign [...] transforms individuals into subjects of a collectively-held history. Its traditional icons, its metaphors, its heroes, its rituals, and its narratives provide an alphabet for a collective consciousness or national subjectivity« (1991: 20).

14 | A central problem for scholars of performance is indeed whether or not Performance Studies constitutes yet another discipline or even an interdisciplinary field that is, in fact, complicit in the reproduction of power/knowledge structures. While to a certain extent this is certainly the case (as the shift from discipline to performance is still underway), Carlson cites two negative responses to this question from the then directors of two major Performance Studies programs in the U.S., Joseph Roach of New York University and Dwight Conquergood of Northwestern University. Roach claims that »[i]t is of course an antidiscipline,« and Conquergood adds that »the trickster [is] the ›guru‹ of this new antidiscipline«. For Carlson himself, it is imperative to recognize that »[p]erformance by its nature resists conclusions, just as it resists the sort of definitions, boundaries, and limits so useful to traditional academic writing and academic structures« (1996/2004: 206).

15 | »Lass mich kurz vorbei, bitte.«

16 | »Wir sind jetzt zwei Wochen da und sind ein bisschen ratlos, weil die Stadt so zerstritten ist. Und wir nicht wissen, wie wir die Leute zusammenbringen sollen; wie wir den Film machen sollen.«

REFERENCES

Anzaldúa, Gloria (1987/1999): *Borderlands/La Frontera: The New Mestiza*, 2nd Edition. San Francisco: Aunt Lute.

Appadurai, Arjun (1996): *Modernity at Large: Cultural Dimensions of Globalization*. Minneapolis: University of Minnesota Press.

Berkhofer, Robert F., Jr. (1979): *The White Man's Indian: Images of the American Indian from Columbus to the Present*. New York: Vintage Books.

Berlant, Lauren (1991): *The Anatomy of National Fantasy: Hawthorne, Utopia, and Everyday Life*. Chicago: University of Chicago Press.

Bhabha, Homi K. (1994/2004): *The Location of Culture*. London: Routledge.

Bird, S. Elizabeth, ed. (1996): *Dressing in Feathers: The Construction of the Indian in American Popular Culture*. Boulder: Westview Press.

Brannen, Mary Yoko (1993): »›Bwana Mickey‹: Constructing Cultural Consumption at Tokyo Disneyland«, in Amy Kaplan/Donald E. Pease (eds.), *Cultures of United States Imperialism*. Durham: Duke University Press, pp. 617–634.

Carlson, Marvin (1996/2004): *Performance: A Critical Introduction*, 2nd Edition. New York: Routledge.

Cracroft, Richard H. (1967): »The American West of Karl May«, *American Quarterly* 19 (2), pp. 249–258.

Deloria, Philip J. (1998): *Playing Indian*. New Haven: Yale University Press.

―――― (2004): *Indians in Unexpected Places*. Lawrence: University Press of Kansas.

―――― (2009): »Broadway and Main: Crossroads, Ghost Roads, and Paths to an American Studies Future«, *American Quarterly* 61 (1), pp. 1–25.

Derrida, Jacques (1993/2006): *Specters of Marx: The State of the Debt, the Work of Mourning, and the New International* (trans. Peggy Kamuf). New York: Routledge.

Desmond, Jane C./Domínguez, Virginia R. (1996): »Resituating American Studies in a Critical Internationalism«, *American Quarterly* 48 (3), pp. 475–490.

Fellner, Astrid M. (2008): »Crossing Borders, Shifting Paradigms: New Perspectives on American Studies«, *Arbeiten aus Anglistik und Amerikanistik* 33 (1), pp. 21–46.

Fishkin, Shelley Fisher (2005): »Crossroads of Cultures: The Transnational Turn in American Studies–Presidential Address to the American Studies Association, November 12, 2004«, *American Quarterly* 57 (1), pp. 17–57.

Foucault, Michel (1975/1995): *Discipline and Punish: The Birth of the Prison*, 2nd Vintage Books Edition. New York: Vintage.

Giles, Paul (2002): *Virtual Americas: Transnational Fictions and the Transatlantic Imaginary*. Durham: Duke University Press.

Gordon, Avery F. (1997/2008): *Ghostly Matters: Haunting and the Sociological Imagination*, 2nd Edition. Minneapolis: University of Minnesota Press.

Halberstam, Judith (2005): *In a Queer Time and Place: Transgender Bodies, Subcultural Lives*. New York: NYU Press.

―――― (2009): »Beyond Broadway and Main: A Response to the Presidential Address«, *American Quarterly* 61 (1), pp. 33–38.

Hardt, Michael/Negri, Antonio (2001): *Empire*. Cambridge: Harvard University Press.

Harvey, David (1990): *The Condition of Postmodernity: An Enquiry into the Origins of Cultural Change*. Malden: Blackwell.

Kaplan, Amy (1993): »›Left Alone with America‹: The Absence of Empire in the Study of American Culture«, in Amy Kaplan/Donald E. Pease (eds.), *Cultures of United States Imperialism*. Durham: Duke University Press, pp. 3–21.

Kaplan, Amy/Pease, Donald E., eds. (1993): *Cultures of United States Imperialism*. Durham: Duke University Press.

Kasson, Joy S. (2000): *Buffalo Bill's Wild West: Celebrity, Memory, and Popular History*. New York: Hill and Wang.

Kooijman, Jaap (2008): *Fabricating the Absolute Fake: America in Contemporary Pop Culture*. Amsterdam: Amsterdam University Press.

Lenz, Günter H. (2002): »Toward a Dialogics of International American Culture Studies: Transnationality, Border Discourses, and Public Culture(s)«, in Donald E. Pease/Robyn Wiegman (eds.), *The Futures of American Studies*. Durham: Duke University Press, pp. 461–485.

Lippert, Leo (2012): »Playing Indian in Austria: Waterloo and the Transnational Performance of America«, in Petra Eckhard/Klaus Rieser/Silvia Schultermandl (eds.), *Contact Spaces of American Culture: Globalizing Local Phenomena*. Vienna: LIT Verlag, pp. 245–265.

Lipsitz, George (1994): *Dangerous Crossroads: Popular Music, Postmodernism and the Poetics of Place*. London: Verso Books.

—— (2001): *American Studies in a Moment of Danger*. Minneapolis: University of Minnesota Press.

Lutz, Hartmut (2002): »German Indianthusiasm: A Socially Constructed German National(ist) Myth«, in Colin G. Galloway/Gerd Gemünden/Susanne Zantop (eds.), *Germans and Indians: Fantasies, Encounters, Projections*. Lincoln: University of Nebraska Press, pp. 167–184.

Maddox, Lucy, ed. (1999): *Locating American Studies: The Evolution of a Discipline*. Baltimore: Johns Hopkins University Press.

McKenzie, Jon. (2001): *Perform or Else: From Discipline to Performance*. London: Routledge.

Moses, L. G. (1996): *Wild West Shows and the Images of American Indians, 1883–1933*. Albuquerque: University of New Mexico Press.

No Name City: Die Freiheit Liegt Südlich von Wien (2006) (AUT, dir. Florian Flicker).

Phelan, Peggy (1993): *Unmarked: The Politics of Performance*. London: Routledge.

Pratt, Mary Louise (1991): »Arts of the Contact Zone«, *Profession* 91, pp. 33–40.

Radway, Janice (1999/2002): »What's in a Name? Presidential Address to the American Studies Association, 20 November 1998«, *American Quarterly* 51 (1), pp. 1–32, rpt. in Donald E. Pease/Robyn Wiegman (eds.), *The Futures of American Studies*. Durham: Duke University Press, pp. 45–75.

Rath, Gudrun (2010): »›Hybridität‹ und ›Dritter Raum‹: Displacements postkolonialer Modelle«, in Eva Eßlinger/Tobias Schlechtriemen/Doris Schweitzer/Alexander Zons (eds.), *Die Figur des Dritten: Ein kulturwissenschaftliches Paradigma*. Berlin: Suhrkamp, pp. 137–149.

Roach, Joseph (1996): *Cities of the Dead: Circum-Atlantic Performance*. New York: Columbia University Press.

Robertson, Roland (1992): *Globalization: Social Theory and Global Culture*. London: Sage.

Schechner, Richard (1985): *Between Theater and Anthropology*. Philadelphia: University of Pennsylvania Press.

Soja, Edward W. (1989): *Postmodern Geographies: The Reassertion of Space in Critical Social Theory*. London: Verso Books.

Taylor, Diana (2003): *The Archive and the Repertoire: Performing Cultural Memory in the Americas*. Durham: Duke University Press.

Traister, Bryce (2010): »The Object of Study; or, Are We Being Transnational Yet?«, *Journal of Transnational American Studies* 2 (1), online.
Wagnleitner, Reinhold (1991/1994): *Coca-Colonization and the Cold War: The Cultural Mission of the United States in Austria after the Second World War* (trans. Diana M. Wolf). Chapel Hill: University of North Carolina Press.
Wikipedia (2011): »*No Name City*«, *Wikipedia* [online], 9 December, http://de.wikipedia.org/wiki/No_Name_City. 15 February 2012.

America, the Threat of Time
Sigmund Skard and Early American Studies

IDA JAHR

> Many of the places one visits around the world attach themselves to memory by their own striking characteristics, by the story they tell, or by the things one has experienced there personally. But some contain, in addition to this, a symbolic value: they are representative of more common aspects of life, which can point the way to the future. Among these latter one finds the Chicago Grain Exchange.
> SIGMUND SKARD, »CHICAGO KORNBØRS«, 1983[1]

In 1947, not long before becoming Norway's first professor of American literature, Sigmund Skard visited Chicago, and he found that the Chicago Grain Exchange represented everything he found disturbing about American civilization. In this 1983 account of the visit, he thus begins his description of the Exchange by reminding his readers of its symbolic value. Skard's travels in 1946 and 1947 took him across the U.S., often by train, and, perfectly exemplifying his attitudes to the city versus the countryside, he continues the short travelogue by describing his approach to Chicago by rail:

I experienced [the Grain Exchange] many years ago, right after the Second World War, at a time when one could still travel by train across the American mainland in a decent fashion. One was then also given time to prepare before arrival. Through many miles, and over many rattling track changes, the plains of Illinois could not seem to decide whether they belonged to the countryside or the city, until finally, the soot-blackened and shabby clusters of houses along the tracks clogged together so that one understood that this had to be the suburbs of the Midwestern Metropolis.[2] (1983: 208)

He then continues by describing the metropolis itself. And, as if to underscore the spatial transformation from rural to industrial landscape across the Illinois plains as echoing a temporal transformation from rural past to industrial future, the city is ›never-ending‹: »The city itself does not present a happier picture, dirty and smoke-swept and never-ending, thunderingly full of noise *on* the streets and *above* them, without characteristics and true center, after Los Angeles the most depressing American example of what humanity can make out of nature«[3] (Skard 1983: 208).

Ambivalences

With his several books on the interpretation of America in Europe, like the massive *American Studies in Europe: Their History and Present Organization* (1958), but perhaps the later and shorter *The American Myth and the European Mind* (1964) in particular, Sigmund Skard earned a reputation as a particularly nuanced critic of American culture and of the European history of looking at America in the international American Studies community. Robert Walker called him a modern Tocqueville (1987: 482),[4] and he is often mentioned as an influential figure in the development of a European American Studies community. In a review of *The American Myth and the European Mind*, Roger Asselineau, one of the founders of French post-war American Studies, points to Skard's balanced view of European myth-making on the one hand and European serious scholarship on American themes on the other (1963: 591). In contrast to this, in their recent survey of Norwegian attitudes to America, *Frykten for Amerika*, Stian Bromark and Dag Herbjørnsrud use Sigmund Skard as an example of prevalent anti-American attitudes in Norway in the post-war era (2003: 158–163), and Norwegianist Stephen Walton places Skard alongside Georges Duhamel as exemplary European males who were challenged by the democratic flatness and openness, what Walton calls ›attention to the gendered (female) characteristics‹, of America (1999: 81–82). This discrepancy—the contrast between his instrumental role in bringing American Studies to Europe and being extolled as a particularly nuanced critic, on the one hand, and his portrayal as a shining example of anti-American sentiment, on the other—is at the heart of understanding Sigmund Skard's work. Indeed, this positioning is also central to an understanding of European attitudes to America, even within—in fact, particularly within—American Studies.[5]

Skard lived for most of the last century, as he was born in 1903 and died in 1995. He was, as indicated above, the first professor of American literature in Norway, accepting the newly formed position at the University of Oslo shortly after World War II. His writing career ran from the early 1930s until the end of the Cold War and his career as an Americanist lasted from the end of World War II until the Vietnam War. When hired, he did not have much knowledge of American literature. In fact, he claimed not to have read a piece of American fiction before he was offered the post (Pells 1997: 96), but he spent the war years in the U.S. and had already worked in the role of explaining America to Norwegians as Chief Regional Specialist for Norway at the Office for War Information in Washington during the war, and there was, literally, no one else who could do the job (cf. Sandved 2002: 25–27; Øverland 2009: 113). Skard was instrumental in forming, and procuring funding for, the European Association of American Studies (EAAS, formed during the Salzburg Conference of 1954) and later the Nordic Association of American Studies (NAAS).

However, »[a]s a European« (Øverland 2009: 119; my emphasis), Sigmund Skard was famously ambivalent toward the United States. In a recent article, Orm Øverland highlights that »this ambivalence, that he describes as ›my own sympathies and antipathies, my deep worries, and my divided heart‹ is a central motif in his memoirs« (2009: 119). That theme was also picked up by Max Silberschmidt, who in a review of Skard's Americanist memoirs *Trans-Atlantica* (1978), comments:

There is a sense of friction, a ›duality‹ underlying the tenor of these Memoirs. This is clearly felt when Skard refers to his giving advice to candidates for US-fellowships by telling them: »Judge America as Norwegians«, which evidently means: Don't get absorbed by a false worship of America, hold on to the ›grim and small nation‹ you belong to. (1979: 87)

In *Trans-Atlantica*, Skard writes that his depictions of America »dramatized his own ambivalence« and notes that some of his worries expressed in the aftermath of World War II may seem »curiously dated« in the late 1970s, »like those of Georges Duhamel 30 years earlier« (1978: 96). However, in *Trans-Atlantica*, Skard still remains highly critical of some aspects of ›American‹ culture, as he »with justification branded many disgusting features of that civilization in its typically American appearance: its senseless confusion, its cultural insipidity and its mechanical and morose opulence« (1978: 96). Here, Skard is decidedly ambiguous, and not just ambivalent. From this text, it is very difficult to discern which parts of his earlier cultural criticism he still stands by. He describes his writings on the Chicago Grain Exchange (with sentences such as »My blood runs cold with fear at the idea that a society thus organized is today going to have its hand at the destiny of the world« [1978: 96]) as one of the texts which likely did not pass the test of time. Yet five years later, he published a short piece on just the Chicago Grain Exchange, describing it as a »Baal temple« and »a temple worthy of the forces it serves,« where the din and cacophony of the »grating human throngs« in the pits is the ultimate image of capitalism (1983: 213; 209; 211).[6]

In the following, I want to explore the contrast described above, the ambivalence or duality mentioned by Øverland and Silberschmidt, not to mention Skard himself, through the metaphor of time, and of timeline—of past, present and future. Sigmund Skard's ambivalence toward the United States' political and economic power after World War II translates into ambiguity in his writings about America. I want to argue that this ambiguity is rooted in Skard placing ›America‹ in two different and hence mutually exclusive points in time simultaneously: America is, rhetorically, at once placed in a pastoral past and a mechanized future. A presentation of Skard's thinking along a specific timeline of development also explains his and other Norwegians' reaction to the perceived threat of American mass culture moving like a wave across Europe after World War II.

Timeline(s)

Throughout modernity, the European academic mindset has tended toward conceiving of time as a linear process. Even though there were other ways of conceptualizing time, these have either been regarded as archaic (like Norse religion) and/or exotic (like Native American religions or Buddhism). Our way of thinking is thus arranged around a notion of progress; if not actual progress, then at least a progression of history, a development. Inherent in the word ›development‹ (as in the German ›Entwicklung‹ and the Norwegian ›utvikling‹) is the notion that there is something to de-velop, to unfold, something already there. This idea of linear time also has its historical predecessors and roots in Aristotelian dramaturgy and Judeo-Christian eschatology, and so, on some level, also includes a move toward an endpoint of history. Depending on one's perspective, this endpoint can be, for example, the fall of the Berlin Wall or the Rapture (which didn't happen, after all, while I was writing this chapter). In the words of Robert Young, the European way of seeing history is »an account of history whose teleology was always directed towards an inexorable closure« (1990/2004: 6).

Post-colonial scholars like Young and Dipesh Chakrabarty have argued since the early 1990s that this particular historicism, this view of history as a linear narrative leading up to now, is at the core of much of Western thinking about globality and globalization. Things happen first in Europe, and then elsewhere:

Historicism enabled European domination of the world in the nineteenth century. Crudely, one might say that it was one important form that the ideology of progress or ›development‹ took from the nineteenth century on. Historicism is what made modernity or capitalism look not simply global but rather as something that became global over time, by originating in one place (Europe) and then spreading outside it. (Chakrabarty 2000: 7)

Europeans have thus placed the (non-white) colonies and later the so-called ›underdeveloped‹ countries along this timeline. In the developmental timeline, Others become predecessors—under-developed. Europeans have long had a tendency to see themselves not just as different but as *better* than their predecessors, and, as Homi Bhabha points out in *The Location of Culture*, Europeans have also had a tendency to find those predecessors in the faces of people who look different (1994/2004: 339–340).

In a commentary in *American Literary History* called »After American Literature«, Peter Carafiol has argued that when Europeans discovered that there was an America, this presented them with a particular problem of history (1992: 539). This new land was not accounted for. And this problem of history prompted a pulling down of the Christian teleological progressive vision from the heavens and applying it to human history, instead »construing an unfamiliar and unaccountable land as a New World, the fulfillment *in potentia* of all the ideals of the Old« (1992: 540). The fulfillment of God's (Manifest) Destiny lay in the inclusion of this New World into History, as God's Plan unfolding on Earth. In different ways and for different reasons, European intellectuals have seen, and portrayed, America as the future of Europe. This, somewhat paradoxically, is the a-historical nature of Enlightenment historicism. In the European imagination, America is not just a place, but also a place in time. Alexis de Tocqueville famously saw more than America in America. In his own words, he saw »the image of democracy itself« but, just as importantly, he saw the future of Europe: a »projected possible historical future« (qtd. in Strout 1969: 87). It is thus hardly surprising that the Norwegian playwright Bjørnstjerne Bjørnson felt the future around him when he visited a women's rally in the States (Skard 1949: 20). Similarly, Swedish novelist and feminist Fredrika Bremer regarded the New World as the heir to the torch of Enlightenment, a spearhead into the bright future of mankind in the 1850s. With the establishment of the American republic, »[t]he community of the United States became the *Mayflower* of the human race« (Bremer 1853: 452).[7] God's Plan for the entire world was manifest in this new continent, with its ideas of freedom for men and women of all religions and races. Progress, or the development of that plan, was inevitable. For Bremer, even slavery was a diamond in the rough; once the institution was ended—and it would be ended—Christian freed slaves would continue God's great work in darkest Africa, spreading the word of mankind's ultimate destiny (cf. 1853: 443–445)—catching up, first to Europe, and then to America.

From the turn of the century onwards, though, and with two successive world wars decimating Europe and its empires, what Arjun Appadurai, echoing Max Weber, calls a view of »the modern world as growing into an iron cage« (1996: 6) was fortified. This led to predictions »that the imagination will be stunted by the forces of commoditization, industrial capitalism and the generalized regimentation and secularization of the world« (1996: 6). By traveling to America, Georges Duhamel, too, traveled to the future: »Humanity still remains so various that it simultaneously offers pictures of an almost paleontological past and living images of the future. Whoever travels in space also travels in history« (1931/1974: xii).

But Georges Duhamel's America was not Frederika Bremer's America. The future was no longer bright and shining. Europeans, among them Knut Hamsun and Weber, and later Duhamel, Skard, and the Frankfurt Institute of Social Research, presented America as a problem to be solved. Duhamel's descriptions in his essays in *Scènes de la Vie Future* (1930; published in English as *America, the Menace: Scenes from the Life of the Future*) place machinistic civilization, mass society, and the fear of automatization at the heart of America. As Jesper Gulddal, for example, points out, European film and literature worked under a metaphorical paradigm which

> describes how the machines in America had become man-eating, accumulating energy for their own functioning via a symbolic and actual devouring of the workers— as it is memorably portrayed in Fritz Lang's classic *Metropolis* (1926), a film which ideologically relies heavily on the European fear of machine civilisation in the United States. (2007: par. 8)

In his inaugural speech in 1948, Skard not just discussed these changes in attitude, but, interestingly, he also echoed them. He set up the two Norwegian writers, Bjørnstjerne Bjørnson and Knut Hamsun, as opposites in their descriptions of America, yet he echoed them both. On the one hand, »America is promises« (1949: 14), while on the other, American civilization peaked some time before the Civil War and the mass influx of catholic immigrants around 1850. The »model democracy in the West« fell prey to »a deep change« which had been going on »behind this façade« (1949: 15–16). Skard claims:

> Up to the 1860's [sic] the immense geographic and industrial expansion in [sic] the North American continent was mostly regarded as a thrilling adventure both by Americans and Europeans, testimony to the power of a young nation. But beginning with the Civil War the transformation became more and more ominous: a new America was taking shape, unrecognizable to the Europeans and even to the Americans themselves. The bold culture of the pioneers seemed to be supplanted by a mechanized mass civilization, dominated by capitalist ›robber barons‹ and a philistine, materialistic middle class. (1949: 16-17)

Everything that seemed to disturb the Turneresque image of antebellum westward expansion toward Manifest Destiny, such as forced removal and slavery, has been curiously excised from the story,[8] and the picture that emerges from this turning point in the plot of American civilization is one of a decline into mob rule and will produce a culture that Skard would later describe as kids gathering with »mindless slack-jawed faces around the record player, rocking to the jungle beat«[9] (qtd. in Gulliksen 1993: 31). And as I have already mentioned above, in his description of the Chicago Grain Exchange,[10] America showed the way to the future.

THE NATION

In one of his autobiographies, Skard writes about how he and three of his brothers built a complete miniature society in and around the family home of Fagerheim, a secluded estate near Kristiansand in the southernmost part of Norway. In his memoir *Solregn* (1980), Skard narrates that the playworld of Salmon River Valley (Laksådalen) was at once a farming village where they »lived the country life in the middle of the city, to Father's great delight« (1980: 22) and a fully formed nation state, with a bureaucracy, kings, governmental bodies, university, and a press. Skard and his brothers tilled the soil and »hammered and joined« (1980: 22), but the brothers also created this society through writing.[11] Sigmund and his brothers wrote the Salmon River Valley into being as a mixture of rural pastoral society, knight's tale, and modern nation state. They wrote the laws and history, bureaucratic and political ordinances, the newspapers (which were sent to elder siblings out in ›the real world‹), and the Valley's national literature. Skard's nostalgia in imagining this make-believe society is evocative of Walter Benjamin's nostalgia in depicting a bygone world in *Berliner Kindheit um 1900* (1932–1934). Benjamin's book is a declaration of love to a city he knows he will have to leave, while Skard's description is of a make-believe nation, a rural, small, grim nation, with representative democracy.

Almost forty years later, in an interview given right after the end of World War II, Skard expressed the opinion that an American scholarly tradition was by then not only a viable but a necessary alternative to the German: »Earlier we had at our higher institutes of learning a strong German attitude, that was both of the good and the bad. Germany is now intellectually paralyzed for some time, so we have to replace it. My plan is to work for a special American institute at the university« (qtd. in C. S. 1946).

Skard's work for American Studies in Europe thus had its pragmatic groundings. But Skard was also able to connect to American Studies because of the field's sense of the nation as the agent of progress. World War II led to a democratic surge among the elites in the U.S. (cf. Gleason 1984: 343;

Ellison 1944/2012). The ideology of democracy received a great revival during the war, partly through the juxtaposition with the barbarism of Nazism, and this opened up a possibility for a linking of democracy and civilization, which should prove invaluable to American Studies pioneers. Philip Gleason has shown how the war, which was seen as a war between cultures, furthered a legitimation of national culture. Pointing to Skard's *American Studies in Europe*, Gleason (1984: 343–344) argues that the growth of American Studies abroad greatly influenced American Studies in the U.S. It was cultural nationalism as a step on the way to forging a common western identity. Skard doubted if a professorship in American Literature was the right position for him, but he had sent a letter to the university to »inquire about his own prospects at the university and in part to point out that the current exclusion of American literature and civilization no longer corresponded ›with our present day realities‹« (Øverland 2009: 112). Though he, for pragmatic reasons specific to the post-war Norwegian university system, was skeptical of interdisciplinarity, Skard was appreciative of the way culture could be read out of literature in the new integrated method (1978: 102). Winfried Fluck has argued that this process of reading culture from a work of art is predicated on a case of metonymy, of letting the parts stand in for the whole. In American Studies, as in all country-specific cultural studies, this, in turn, assumes a whole—an organic national culture (1978/2009: 33). Sigmund Skard could find a home in American Studies precisely because of this assumption of a national culture, which was formed into a perfected unity over time. In 1948, according to Skard, »[i]n many ways the United States [was] still little more than the beginning of a nation, distorted by the spasms of fierce contrasts,« but it was bound to grow into its own, and achieve »that depth and perspective, that variety of shades which is the privilege of established societies« (1949: 9).

By viewing the nation as the locus of progress, Skard placed himself in a long line of Norwegian literary historians. In 1848, the time of failed revolutions and of Fredrika Bremer's trip to see the ›Homes in the New World‹, linguist Ivar Aasen collected dialect samples from different parts of Norway and created a language he called Landsmål (Country Tongue), but which later was renamed to *Nynorsk*, or New Norse. At the time, the bureaucratic elites in Norway would speak and write Danish, while the ›lower‹ (which in Norway at that time meant small farmer, rather than industrial working) classes spoke Norwegian dialects. This New Norse, ›the people's language‹, was to become one of the most important causes for Sigmund Skard, as it was for his father, and his parents-in-law, as well as for his wife.[12] The Norwegian Anglicist Arthur Sandved points out that Skard did, indeed, write very little on American literature, but was much more active in the fight for the increased use of Nynorsk in written language (2002: 237). Yet, this is not entirely true. As his hand-written lectures on American authors, which are available in archives of the Norwegian National

Library, demonstrate, Skard wrote a lot on American literature, he simply never published much of it.

However, Skard was, indeed, very active in the New Norse language movement (in Norwegian referred to as the ›language rising‹) and it is as a central figure in this movement—and as a poet—that he is best known in his home country. Skard was a vocal advocate for New Norse in Norwegian newspapers and on the radio. In addition, he was chairman of the board of the largest New Norse publishing house in Norway for most of his career. He published more than fifteen books, translations, and original works with this publishing house. In a recent article on Skard, Orm Øverland describes Skard as »a fine scholar who put his own work on hold during the decades he devoted himself to his foundational work for a new field of academic study« (2009: 120), indicating that American Studies was not Skard's *own work*. Sigmund Skard's own work, as it is alluded to by Øverland, was with the language movement, among other things. Øyvind Gulliksen has suggested that Skard's work in American Studies should be considered in light of his work with the nationalist language movement: »When [Skard] met typical British attitudes, like that American literature was simply imitative, ›derivative‹, then Skard in a way reacted in *Nynorsk*«[13] (1993: 29). Thus, Sigmund Skard saw parallels between the fight for New Norse against the Danish language and American culture, literature, and language standing up to old British imperialist attitudes.[14] This might be the reason why so much of his writing about America is infused with a deep-rooted skepticism of industrialization and with a nostalgic yearning for country life and the times of the early settlers.[15]

In Skard's writing, the function of the trope ›America‹ is thus at times to strengthen and at times to offset Norwegian national identity. At times, Skard uses ›America‹ as an image portraying his concept of the nation—the pinnacle of progress and a step toward a bright future. America is squarely placed in the past, in a romanticized, narrated past that exists somewhere around the middle of the nineteenth century. In keeping with his contemporary American Studies tradition, in many of his writings, ›America‹ is located in the time of the American Renaissance and the westward expansion, because this opens up for a rural America, an America that Skard can compare with rural Norwegian traditions he sees as the well from which democracy again shall flow. At other times, ›America‹ is the image of a modernity which scared him deeply, as in his writings on the Chicago Grain Exchange or on the leveling effects of the spread of suburbs in the U.S. In the latter case, Skard was acting as what Stephen Walton has called a ›seer‹, an intellectual of high modernity who writes of ›America‹ as the future of the world (1999: 86).[16] Skard's work on America is ambiguous because his ›America‹ is imagined in two different and mutually exclusive points in time at once, and he is unable to reconcile the two. The organic American national culture he can relate to through American literary studies at the time exists sometime around 1850, and the America described in much of his writings for a Norwegian audience exists, like Duhamel's, in the future. These are, in fact, irreconcilable points in time along the developmental timeline, and Skard's narrative cannot sustain them both. This is what gives rise not just to the aforementioned ambivalence in Skard's relationship with America, but also to a curious ambiguity in all of his writings.

Global time

The concept of a chronological and linear timeline and the ruptures in this timeline that appear in Skard trying to reconcile two different Americas necessary to perform his professorial duties is also fruitful for thinking about the globalization processes mentioned earlier in this chapter. The nation is an important concept and issue for post-colonial theorists. Homi Bhabha, for instance, derives his sense of nationalism, at least in early writings, from Benedict Anderson, and the ›imagined community‹ is an important metaphorical construction for understanding Skard's relationship to America, as well. Skard's nation, his Europe, or his ›classical tradition‹ for that matter, are all imagined communities, and the sense of the interconnectedness of the nation and the growth of representative democracy—both arising from print culture and its concomitant culture of mass education—is an insight I find important. But for the purposes of this chapter, I consider it just as useful to employ sociologist Ernest Gellner's sense of the national, which is first and foremost a political concept. Gellner defines nationalism as *congruence between state and ethnicity*; as a political principle, and, importantly, nationalist sentiment and nationalist movements flow from this political principle being violated in some way (1983/2006: 2). Gellner's definition is valuable, for it allows explaining both of Sigmund Skard's reactions to America from a nationalist perspective. Skard felt he could relate to American Studies because the field is inherently interconnected with the nation as a representative unit. However, at the same time, transnational capitalist forces, which he saw as emanating from America like a giant capitalist octopus with tentacles reaching all over the world, as described in »The Chicago Grain Exchange«, are threatening to disturb his attempts at creating a nation state that will be truly representative of the Norwegian people.

The rupture in the narrative of teleological historicism appears when the future is suddenly represented by something unrecognizable. As Skard argues in *The American Myth*:

The magnitude of the westward movement and typical life along the frontier were strange to European eyes. But the civilization which developed largely appeared recognizable; somehow the inherited set of Old World standards could still be applied. After the Civil War, this was no longer true to the same extent. The new form of life which developed in America was characterized by a speed and dimension, a violence and recklessness which had no exact correspondence even in industrialized Europe. (1964: 35)

And when the unrecognizable impacts daily life, when Europe no longer seems to have the guiding role in the progressive fairy tale, and is no longer the end result in the narrative, when Europeans are at the mercy of forces over which they have no control, this creates a disruption in the timeline; the ambiguity in Skard's writing. In *Målstrid og Massekultur*, Skard describes American postwar life thus:

It is the total consumer society which here is being formed before our eyes. It builds on an economic cooperation of capital and work, technical research and public efforts more thorough than the world has ever seen. The veins of the new society are the large automobile roads across the country, which now carry up to 95 percent of the traffic. Along these roads people flock to the cities, and stay there; only 6 percent of America's population remains today in the country as independent farmers. But more and more these people only work in the cities. They live in the suburbs, sleep-towns, which in a few years will have transformed all of the American east and west coast into a continuous belt of villas.

Here is the home of the holy and common middle class, which soon will have sucked up the other classes both from above and below and instead is molded into an even more standardized all-American civilization.[17] (1963: 8-9)

Skard retired from American Studies in 1973. Although he continued writing and publishing extensively until the late 1980s, very little of what he wrote was on American themes. In a letter to the American literature professor Merle Curti, Skard wrote that although he himself was quitting, he did not see American Studies as any less important. On the contrary, he claims, American Studies is »even more [important,] the more America shows its real character as a testing ground of the development of the entire world« (1972). And this development »scare[d] [him] more than [he] c[ould] say« (Skard 1972).

A timeline of development implies a past, present, and a future. As long as the future is us, we are perfectly happy with the idea. However, when the future is someone else, and someone, in Skard's terminology, unrecognizable, we are not. The progressive myth of enlightenment incorporated the representative nation state as one of its central features, and the seeming loss of control over the powers of capital and representation to transnational capitalist forces created the extreme fear already visible in *The Study of American Literature* and the 1949 book *Amerikanske Problem*. However, for Skard, both the hope for the perfected nation (democratic, not fascist) and the threatening transnational forces emanate from the same place. If we are to read his attitudes to the new powerhouse on the block from his inaugural speech in 1948, we can see his wavering. Here, the dark clouds are tangible, but contained. In the talk, Skard focuses on creating a common tradition between Norway and the United States. Even though he juxtaposes the positive Bjørnstjerne Bjørnson with the very negative Knut Hamsun, he seems to echo them both, as he creates a narrative arc of progress and decay throughout American history, with the zenith located around the time of mass immigration and industrialization. The pinnacle of American civilization, according to this speech, was the end of the nineteenth century. The progressives did what they could with little to work on, and since then, it has all been going downhill. Sigmund Skard was a man who saw the rubbish pile mount at the feet of Walter Benjamin's angel of history and the winds of progress that took hold of its wings, propelling it backward into the future. For a while, American Studies seemed a large sturdy tree of nationalism he could hold on to in order not to be swept up by those same winds, but eventually, he saw that the tree was a branch—was a twig—and it broke.

Notes

1 | »Mange av dei stadene ein har gjesta kring i verda, bit seg fast i minnet ved sine eigne slåande særmerke, ved den historia dei fortel om, eller ved slikt som ein sjølv har opplevt der. Men somme har attåt dette ein suymbolverdi: dei står for meir allmenne livsdrag, som også kan peika inn I framtida. Mellom desse siste høyrer kornbørsen i Chicago.« All translations from Norwegian are mine.

2 | »Eg opplevde den for aldri så mange år sidan, like etter den andre verdskrigen, den gongen då ein enno køyrde sømeleg med tog over det amerikanske fastlandet. Da fekk ein òg stunder til å bu seg før ein var framme. Gjennom aldri så mange miler og over aldri så mange raslande sporskifte kunne slettene I Illinois liksom ikkje bli samde med seg sjølve om dei skulle vera land eller by, til endeleg dei nedsota og loslitne husklynjene langsmed sporet klabba seg slik I hop at ein skjøna det laut vera forstadene til Metropolis i Midtvesten.«

3 | »Byen sjølv tar seg ikkje gladare ut, skiten og røyksveipt og endelaus, dundrande bråksam på gatene og over dei, utan særdrag og verkeleg sentrum, nest Los Angeles det tristaste amerikanske dømet på kva menneske kan laga ut av naturen.«

4 | In the review, Walker chimes in with Lewis Hanke's description of Skard as »a modern Tocqueville«, before he chides Hanke for preferring the Norwegian version of Skard's *The United States in Norwegian History*, wondering if it is because it is in Norwegian or because the English language version is published in the U.S. In reality, the Norwegian version of the book is considerably longer and with an array of notes which is absent from the English language publication in Walker's Contributions in American Studies series.

5 | Sigmund Skard's American Studies was the study of America within the then-established disciplines. Hence, in his books on the subject, he includes work done in history, literature, and geography, among others. Back then, the American Studies community was only just beginning to see itself as something other than a branch of English literature. However, there was already a strong sense of the book as a national issue. Today, ›American Studies‹ is a term that can almost, but not quite, be described as a floating signifier. There is a ›relational geography‹ within the American Studies community (cf. Hones/Leyda 2005), but it is one that overlaps with several other fields and disciplines and several other discourses. So, when I write about Sigmund Skard being interesting within the context of, or the history of—and perhaps even to understand—American Studies in Europe, I am thinking of something which perhaps has to be triangulated by referring to the conferences, the journals, and the academic exchange programs that make up this relational geography, but which also is embedded in a very specific set of historical circumstances, as the conferences, journals, publication series, and academic exchange programs are historically embedded, and they determine the nature of American Studies as performed.

6 | The full sentence on page 209 reads : »Det er heller ikkje noko vanleg hus, men eit temple, verdig dei makter det tener.« On page 211, Skard uses the formulation »samanelta folkehopane med sin synlege lidenskap, med kvite andlet, vilt svingande armar og peikande hender.«

7 | When reading this part of Bremer's letters, I picture John Gast's *American Progress* (ca. 1872), her white, almost shining, gown, gliding across the continent. I know that's not what Bremer looked like, but one can dream.

8 | Still, the race ›problem‹ was one of Skard's favorite issues, on which he gave numerous speeches, most of them strongly influenced by Gunnar Myrdal's *An American Dilemma*. Traces of Skard's lecturing activities after the war can be found in both the archives of the newspaper *Verdens Gang* and of the NRK, the Norwegian Public Broadcasting Corporation.

9 | The full quote reads: »Den nye innstillinga har fått sitt menneskjelege symbol i dei flokkane av reint ung ungdom som vi alle har sett, sitjande på huk med tankesløkte fjes kring platespelaren eller transistoren innstilt på Radio Lux, og ruggande i urskogstakt, ikkje etter artistisk jazzmusikk, men etter den siste såpeglatte slager, som ikkje lenger er eit tidtrøyte i mellomstunder, men det viktigaste åndsinnhaldet.«

10 | This text was published in *Vandringar* in 1983, but I have found it to be heavily based on notes written as early as 1947.

11 | »Der drev vi jordbruk med reddikar og buskar, tømra og snikra, laga mat på den ørvesle peisen, og levde bondeliv midt i byen, til fars store fryd.«

12 | In Norway, Skard's wife, Åse Gruda Skard, is famous as the person who introduced child psychology to the public, and ›the grandmother of all Norwegian children‹. Her father was Professor of History and for a time Foreign Minister in Norway.

13 | »Når han møtte typiske britiske holdninger om at amerikansk litteratur bare var etterplapring, ›derivative‹, så reagerte Skard på sett og vis på nynorsk.«

14 | Skard himself drew this parallel in a letter he wrote to his old professor Francis Bull (cf. Skard 1945).

15 | Though it regarded itself as a popular movement, the Norwegian Nynorsk movement was never able to connect with the rising industrial labor class in Norway, though Skard's father-in-law, Halvdan Koht, was one of those who tried (cf. Angell 2001: 84-89).
16 | Other seers in Scandinavia at the time, according to Walton, were Skard's wife, Åse Gruda Skard, and the couple Gunnar and Alva Myrdal. Walton also compares the two Scandinavian couples with Sartre and de Beauvoir in France.
17 | »Det er det totale forbrukersamfunnet som her blir til for augo våre, kommersialisert, mekanisert, automatisert og uniformert. Det byggjer på ei økonomisk samordning av kapital og arbeid, teknisk granskning og offentlege tiltak som er meir gjennomført enn verda nokon gong før har sett. Blodårene I det nye samfunnet er dei store bilvegane tvert over fastlandet, som no tar bort imot 95 prosent av trafikken. Etter desse vegane strøymer folket til byane , og blir der; bere 6 prosent av USA's folkemengd er I dag att på bygdene som sjølveigande bønder. Men meir og meir har desse bymenneska berre arbeidet sitt i byane. Dei lever i forstadene, sovebyane, som om få år vil ha gjort heile den amerikanske aust- og vestkysten om til samanhengande villabelte. Her er heimen for den heilage og almenne mellomklassa, som snart vil ha soge den andre klassane opp i seg både ovantil og nedantil og I staden sjølv er støypt inn i ein alt meir einsarta, all-amerikansk sivilisasjon.«

References

Anderson, Benedict (1991): *Imagined Communities: Reflections on the Origin and Spread of Nationalism*. London: Verso Books.

Angell, Svein Ivar (2001): »Språkspørsmålet som en del av den norske moderniseringsprosessen«, in Elisabeth Bakke/Teigen Håvard Teigen (eds.), *Kampen for språket: nynorsken mellom det lokale og det globale*. Oslo: Det Norske Samlaget, pp. 72–91.

Appadurai, Arjun (1996): *Modernity at Large: Cultural Dimensions of Globalization*. Minneapolis: University of Minnesota Press.

Asselineau, Roger (1963): »[Untitled Review of Sigmund Skard's *The American Myth and the European Mind: American Studies in Europe, 1776–1960*]«, *American Literature* 34 (4), pp. 591–592.

Benjamin, Walter (1932–1934/2006): *Berlin Childhood around 1900* (trans. Howard Eiland). Cambridge: Belknap Press.

Bhabha, Homi K. (1994/2004): *The Location of Culture*. London: Routledge.

Bremer, Fredrika (1853): »Letter XXXVII: Letter to Her Majesty, Carolina Amelia, Queen Dowager of Denmark« (trans. Mary Botham Howitt), in *Homes of the New World: Impressions of America*. New York: Harper, pp. 420-452.

Bromark, Stian/Herbjørnsrud, Dag (2003): *Frykten for Amerika: en europeisk historie*. Oslo: Tiden Norsk Forlag.

C. S. (1946): »Plan om et særskilt amerikansk institutt ved Oslo universitetet«, *Nordisk Tidende*, 12 September, p. 16.

Carafiol, Peter (1992): »Commentary: After American Literature«, *American Literary History* 4 (3), pp. 539–549.

Chakrabarty, Dipesh (2000): *Provincializing Europe: Postcolonial Thought and Historical Difference*. Princeton: Princeton University Press.

Duhamel, Georges (1931/1974): *America, the Menace: Scenes from the Life of the Future* (trans. Charles Miner Thompson). New York: Arno Press.

Ellison, Ralph (1944/2012): »*An American Dilemma*: A Review«, *Teaching American History* [online], n. d., http://teachingamericanhistory.org/library/index.asp?document=554. 7 May 2011.

Fluck, Winfried (1978/2009): »Aesthetic Premises in American Studies«, in Robin W. Winks (ed.), *Other Voices, Other Views: An International Collection of Essays from the Bicentennial*. Westport: Greenwood Press, pp. 21–30, rpt. in Laura Bieger/Johannes Völz (eds.), *Romance with America? Essays on Culture, Literature and American Studies*. Heidelberg: Universitätsverlag Winter, pp. 15–38.

Gellner, Ernest (1983/ 2006): *Nations and Nationalism*, 2nd Edition. Ithaca: Cornell University Press.

Gleason, Philip (1984): »World War II and the Development of American Studies«, *American Quarterly* 36 (3), pp. 343–358.

Gulddal, Jesper (2007): »A Heavy Prelude to Chaos: Aspects of Literary Anti-Americanism in the Interwar Years«, *Eurozine* [online], 20 March, http://www.eurozine.com/articles/2007-03-20-gulddal-en.html. 7 May 2011.

Gulliksen, Øyvind T. (1993): »Sigmund Skard: amerikanist på nynorsk«, *Mål og makt* 4, pp. 23–41.

Hones, Sheila/Leyda, Julia (2005): »Geographies of American Studies«, *American Quarterly* 57 (4), pp. 1019–1032.

Øverland, Orm (2009): »Getting Started: Comparative Notes on the Impact of Sigmund Skard on American Studies in Norway«, *American Studies in Scandinavia* 41 (1), pp. 108–121.

Pells, Richard (1997): *Not Like Us: How Europeans Have Loved, Hated, and Transformed American Culture Since World War II*. New York: Basic Books.

Sandved, Arthur O. (2002): *Fra ›kremmersprog‹ til verdensspråk, 2, Engelskfaget ved Universitet i Oslo 1945–1957*. Oslo: Forum for universitetshistorie.

Silberschmidt, Max (1979): »[Untitled Review of Sigmund Skard's *Trans-Atlantica: Memoirs of a Norwegian Americanist*]«, *American Studies in Scandinavia* 11 (2), pp. 86–87.

Skard, Sigmund (1945): »Letter to Francis Bull, June 18th, 1945«, *Amerikansk Institutt organisasjonshistorie*. Oslo: University of Oslo, Faculty of the Humanities Archives, Folder »Utlånt fra Amerikansk Institutt«.

―― (1949): *The Study of American Literature*. Philadelphia: University of Pennsylvania Press.

―― (1958): *American Studies in Europe: Their History and Present Organization*, 2 Volumes. Philadelphia: University of Pennsylvania Press.

―― (1963): *Målstrid og massekultur: tankar til ettertanke*. Oslo: Det Norske Samlaget.

―― (1964): *The American Myth and the European Mind: American Studies in Europe, 1776–1960*. New York: Barnes.

―― (1972): »Letter to Merle Curti, July 9th, 1972«, *Letters to Sigmund Skard*. Oslo: Norwegian National Library Letter Collection, Folder 641.

―― (1978): *Trans-Atlantica: Memoirs of a Norwegian Americanist*. Oslo: Universitetsforlaget.

―― (1980): *Solregn: ein sjølvbiografi*. Oslo: Samlaget.

―― (1983): »Chicago Kornbørs«, in *Vandringar*. Oslo: Det Norske Samlaget, pp. 208–214.

Strout, Cushing (1969): »Tocqueville's Duality: Describing America and Thinking of Europe«, *American Quarterly* 21 (1), pp. 87–99.

Walker, Robert (1987): »A World of Tocquevilles?«, *American Quarterly* 39 (3), pp. 480–484.

Walton, Stephen J. (1999): »Blood, Piss and Shit: The Challenge of America to the Male European Imagination«, in *Litt laust, mest fast*. Volda: Høgskulen i Volda, pp. 80–93.

Young, Robert J.C. (1990/2004): *White Mythologies: Writing History and the West*, 2nd Edition. London: Routledge.

Real Places and Imaginary Spaces

Setting the Scene
L. M. Montgomery's Imaginative Island Landscapes

JULIA VAN LILL

Spaces and places play an important role in literature for children and adolescents. Indeed, some of the most memorable settings in books come from fiction for children or young adults: Narnia, Never-Land, or Wonderland, to name but a few. One of the reasons for this is that these spaces are closely associated with juvenile imagination and reading about fantasy landscapes enables young readers to transcend their ordinary surroundings and give free reign to their creativity.

However, the trope of fantasy landscapes and alternate spaces in fiction for young readers is a relatively new one. Prior to the nineteenth century, literature for children and adolescents was written mainly for the purpose of moral and religious instruction. Unsurprisingly, setting was only of secondary importance. It was only with the advent of the Romantics that the general perception of children began to change and the imagination gained meaning and status. Children were no longer seen as bearers of original sin with a need for strict education and guidance, but as innocent and in need of protection. In addition, the Romantics began to equate imagination with the ability to »make familiar objects be as if they were not familiar« (Shelley 1840/2009: 353), an ability that for Samuel Taylor Coleridge is similar to »the child's sense of wonder and novelty« (1802/1996: 80).

With this new view of children, books written for young readers began to change. Although morals were still an important part of children's fiction, entertainment and the encouragement of imagination soon became its main purpose. By the mid-nineteenth century, authors for the young reading public had given free reign to their imagination, and authors such as Edward Lear, Charles Kingsley, and Lewis Carroll became leading writers in the market for children's literature. Even writers of more realistic children's books took on characteristics of fantasy or fairy fiction. Magical coincidences, imaginative protagonists, or fairy tale landscapes were only some of the ways in which non-fantasy writers for children and young adults such as Frances Hodgson Burnett or Kate Douglas Wiggin incorporated make-believe and imaginative creation into their works.

Another such writer was Canadian-born Lucy Maud Montgomery, whose books abound in imaginative heroines traipsing through fantasy-esque landscapes. To this day, Montgomery remains one of the most important writers of young girls' fiction. Her first girls' novel, *Anne of Green Gables*, was published in 1908 and was such a success that publishers and readers quickly demanded a sequel, which Montgomery wrote and published only a year later, followed by six further novels in the *Anne* series. Between her initial success with *Anne of Green Gables* and her death in 1939, Montgomery wrote a total of twenty novels and numerous short stories, some of which were published only post-humously. A large part of Montgomery's enduring appeal are her vivacious protagonists and charming narrative style, but also her artistic and fanciful Prince Edward Island (PEI) landscapes in which she set the majority of her stories and novels. When reading Montgomery's novels, it becomes clear that Prince Edward Island held a special place in her heart and was a contributing source of her own creativity. Her novels show both the impact of Montgomery's own life and surroundings, as well as the Romantic influence on views of childhood, nature, and imagination.

Nature and imagination were intimately connected for Montgomery, which is indicated by her description of »creative realm of imagination and fantasy« as her »secret garden« (Gammel 2002b: 116). ›Secret garden‹ implies a private space not only intimately connected with nature, but it lends an organic subtext to the concept of imaginative creation, as if thoughts and ideas grew in a similar way to flowers, shrubs, and trees. In *Emily of New Moon*, this idea is referred to as »a fertile invention« (Montgomery 1923/1993: 118). The term also indicates the need for nurturing, however. Unlike a sublime wilderness landscape such as the Alps, a garden must be tended and cultivated in order to grow. Flowers must be planted and weeds removed for the garden space to achieve its full potential. Ugly thoughts have no place in a beautiful creative arena. Montgomery dramatizes this idea in *Anne of Green Gables* in the chapter entitled »A Good Imagination Gone Wrong«. In this chapter, Anne and her best friend Diana Barry create a series of ghosts and goblins that supposedly inhabit the grove behind Green Gables farm. When asked to run an errand one evening, Anne's ugly imagination overtakes her good sense and she lets the monsters that she and Diana created overpower her, that is, she faints, and the beauty of the grove is spoilt for her forever. Kathleen Ann Miller comments on Anne's imaginative mishap in creating the Haunted Wood: »If Anne's imagination can control and shape the landscape [by peopling it with ghosts and goblins], then her imagination must be trained for good use« (2010: 134). Like the garden, her creative powers must be weeded and trimmed into shape.

The garden space was a significant setting in writing for children from the late eighteenth- and early nineteenth-century onwards, particularly girls' fiction, reaching its most prominent exemple later on in 1911 in Burnett's *The Secret Garden*. Elsie L. Smith investigates the uses of gardens, arbors, and bowers in her article »Centering the Home-Garden: The Arbor, Wall, and Gate in Moral Tales for Children,« in which she states that »[t]he cultivated garden was seen as a training ground for children—set apart from the wilderness of nature but more flexible, both spatially and socially, than the confined rooms of the home« (2008: 24). Kerry Mallan makes a similar observation in her discussion of Australian picture books, arguing that the garden or backyard

provides a particular kind of playspace that suggests a degree of independence, retreat, and ownership, with a limitless potential for make-belief. Though it is located within the domestic domain and its perimeters are often bounded by a fence, the backyard still offers a space for freedom of movement and covert play. (2003: 167)

Both of these critics suggest an innate connection between the garden space and childhood. The garden also creates the potential for growth, and physical and imaginative freedom. Montgomery's use of a garden metaphor for her creative mind thus implies a connection between her imagination and childhood, as well as a localized aspect of her creativity centered on her immediate (domestic) surroundings. It is perhaps not surprising that Montgomery's fiction is almost exclusively set on her native soil of Prince Edward Island, Canada; her extended backyard so to speak. Some of the most memorable locales in her novels are based on real places from her childhood, such as Lovers' Lane or the Haunted Wood. Given the connection between imagination and nature that Montgomery made in her journals, the spatial settings in her novels deserve closer consideration for they forge (inter)connections between imagination and adolescence.

Several critics, Elizabeth Rollins Epperly, Irene Gammel, Janice Fiamengo, and Marylin Solt among them, have commented on the uses of setting and landscape in Montgomery's fiction. In her article »Embodied Landscape Aesthetics in *Anne of Green Gables*,« Gammel comments on the extravagant descriptions of nature in Montgomery's most popular novel, *Anne of Green Gables*. She likens the author's portrayal of landscapes to that of her Romantic predecessors, the poets that Montgomery so loved to read, noting the vivid »painterly« (2010: 229) quality that informs the novel's nature descriptions. However, Gammel also notes that Montgomery relies not only on the visual in her descriptions of nature but furthermore employs scent, sound, and sensation in order to immerse the readers and her heroine more fully within the landscape. By doing so, she argues, Montgomery »blurs the boundaries between inside and outside, life and art, human and nature, physical and metaphysical« (2010: 229). This »multi-sensuous aesthetic« thus transforms the landscape into a »sensorial training ground for Anne's developing identity« (2010: 230). Gammel's argument points to an aspect of duality in Montgomery's settings, a fusion of external nature and the internal life of the protagonist. This concept was not a new one in nineteenth-century women's writing. Charlotte Brontë's desolate moors in *Jane Eyre* (1847), to give but one example, vividly mirror Jane's own wretched anguish. However, given Montgomery's comparison of nature with her creative space in her journals and her reverence of Romantic poetry that so often connected imagination with nature, I would like to argue that rather than connecting her landscape settings with her protagonist's identities, nature and setting in the *Anne* and *Emily* novels function as the landscapes of fantasy and imagination and are intimately connected with her young protagonists' creativity and inner mental space.

Although based largely on the scenery of her Cavendish home, Montgomery's Avonlea settings in the *Anne of Green Gables* books nevertheless convey a sense of mysticism and fantasy that negate the aspects of more rigid realism or regionalism in her writing. Montgomery's landscapes transcend their real-life models and hold a deeper meaning than merely the setting of a story.

In the introduction to *Making Avonlea: L. M. Montgomery and Popular Culture*, Irene Gammel points to the enchantment that even at first glance surrounds the village of Avonlea:

›Avonlea,‹ Montgomery's name for Cavendish, is at the border of the real and the fictional, the regional and the universal. Avonlea has a mythopoetic quality, the name derived from a rich combination of Arthurian legend (Avalon = the enchanted island), classical myth (*Insula Avallonis* = Isle of Apples), and Shakespearean legend (Stratford-upon-Avon), with Avon-lane suggesting the entrance into Montgomery's realm. (2002a: 10-11)

Already the name of the village points to its unique quality and a hint of romance and fantasy. Unlike the other towns and villages in the Anne books, Charlottetown or White Sands, for example, whose names lack the mysticism suggested by Avonlea, Anne's new hometown is set apart as a place on the border to an imaginary land and seems a most appropriate home for the novel's highly imaginative heroine. Much of this enchantment is conveyed by the descriptions of nature. Montgomery uses a wealth of vivid adjectives and visual suggestion to bring her Avonlea landscapes to life for the reader. She uses a multitude of colors, gemstones, flowers, and metaphors in order to achieve a setting that resembles a brilliant Edenic garden. Elizabeth Epperly calls this type of writing ›prose-poetry‹ (1992: 28). Examples of such descriptions are »horizon mists of pearl and purple«, »the light came down sifted through so many emerald screens that it was as flawless as the heart of a diamond,« and »the sea, misty and purple, with its haunting, unceasing murmur« (Montgomery 1908/1992: 9; 106; 306). What makes these descriptions so magical is Montgomery's combination of extravagant imagery, such as the pearl and diamond images, infusing the natural world with a sense of wealth, and a status above the ›normal‹ farmland that it actually was, and the Romantic idealization implied in its elusive perfection.

Above all, however, Montgomery achieves her blurring of reality and fantasy by using images of vagueness and obscurity. Her landscapes abound in mists and hazes, lights and shadows, and colors that shift and blur. For example, when Anne first sees Barry's pond on her way to Avonlea, readers are told that »the water was a glory of many shifting hues—the most spiritual shadings of crocus and rose and ethereal green, with other elusive tintings for which no name has ever been found« (Montgomery 1908/1992: 19). The words ›shifting‹, ›shadings‹, ›ethereal‹, and ›elusive‹ within this brief portrayal of the pond all convey a sense of magical mystery by creating a dream-like atmosphere, both vague and intangible. This vagueness of setting reminds readers strongly of New Woman writer Kate Chopin's *The Awakening* (1899), in which the scenes, particularly those on the island of Chênière Caminada, are similarly clouded by dreamy hazes and an elusive atmosphere. The phrase »misty spirit forms were prowling in the shadows and among the reeds, and upon the water were phantom ships, speeding to cover« (Chopin 1899/1993: 39) could almost be a passage from Montgomery's novel. Chopin employs these dream scenes in order to underline her protagonist Edna Pontellier's state of in-betweenness; she is neither angel nor demon, neither asleep nor awake, and she is at home neither on the sexually more open holiday island of Grand Isle nor in the comparatively restricted, structured city of New Orleans her marital home.

But where Chopin's heroine embodies the difficulties of turn-of-the-century women, Anne Shirley's in-betweenness results from her situation of being neither child nor adult, she is somewhere in that vague space of adolescence. She is also an orphan, a person without a home or roots. Before arriving at Green Gables, Anne was quite literally a wanderer, moving from one foster home to the next. But Anne soon feels at home at Green Gables and forces her individuality on her new surroundings. Upon seeing the lovely pond on the Barry lands, Anne promptly names it the ›Lake of Shining Waters‹. Fiamengo argues that Anne, by finding new titles for the places she loves, »names the landscape into her power, lovingly but authoritatively« (2002: 232). Anne continues to name and rename most of the landscape and nature around her and subsequently all of these places are known by their ›Anne-names‹. Readers rarely recall Barry's pond but will always remember the Lake of Shining Waters. This has the effect that Avonlea takes on the essence of Anne's imagination, making it a direct reflection of Anne's creativity. Through Anne, rural Avonlea becomes an enchanted country whose map would read something like ›north of the Lake of Shining Waters, through Lovers' Lane and past the Haunted Wood.‹ The Prince Edward Island of Anne's imagination transcends its farmland designation and turns into a fairy tale landscape perfect for adventures of all kinds.

In contrast to the magical, Romantic landscape descriptions of Prince Edward Island stands the sensible, no-nonsense Cuthbert farmstead, Green Gables. Readers are first introduced to Anne's new home via the backyard. It is described as »[v]ery green and neat and precise [...], set about on one side with great patriarchal willows and on the other side with prim Lombardies« (Montgomery 1908/1992: 3–4). This backyard is the complete opposite of the dream-like nature of the island. It is rigid and unimaginative and reflects its two elderly sibling residents: ›patriarchal‹ man-of-the-house Matthew and ›prim‹ Marilla. Nowhere is there (yet) a trace of the young girl Anne who is about to descend upon them. Neighbor Rachel Lynde mentions that »Marilla Cuthbert swept that yard over as often as she swept her house« (1908/1992: 4). This statement points to the trope of the backyard as a clear extension of the house, to be kept as neat and orderly as the parlor. Nevertheless, there are certain redeeming qualities on the Cuthbert farm that allow for more ›scope for imagination‹. Through the east window of the kitchen, »you g[e]t a glimpse of the bloom of white cherry trees in the left orchard and nodding, slender birches down in the hollow by the brook« (1908/1992: 4). These are the things that Anne feels a certain kinship with from the very beginning. Before she even sets foot outside the Green Gables door, she states »I want to go out so much—every-thing seems to be calling to me, ›Anne, Anne, come out to us. Anne, Anne, we want a playmate‹« (1908/1992: 34). Anne seems to commune with the nature around her.

Shortly after this initial introduction to Green Gables, the farm is depicted through the eyes of Anne Shirley. Anne, scanning the countryside for the unknown house that is to be her home, looks from one homestead to the next until »[a]t last [her eyes] lingered on one away to the left, far back from the road, dimly white with blossoming trees in the twilight of the surrounding woods. Over it, in the stainless southwest sky, a great crystal-white star was shining like a lamp of guidance and promise« (1908/1992: 21). Anne comments that »just as soon as I saw it I felt it was home« (1908/1992: 21). The shining star is reminiscent of the biblical star that guided the wise men to the manger and the new-born Christ, indicating in its mysticism that Green Gables is to be the salvation of the poor, red-headed orphan girl.

After she has been officially installed as a resident of Green Gables, Anne begins to affect little changes to the house that stake her territorial claim. She names the trees and flowers in and around the house, decorates her room and others with leaves and flowers from outside, and suffuses the house with an atmosphere of liveliness and laughter. It is particularly her little gable room that shows the change induced by her imaginative personality. Toward the end of the novel, »[t]he east gable was a very different place from what it had been on that night four years before, [...] Changes had crept in, Marilla conniving at them resignedly, until it was as sweet and dainty a nest as a young girl could desire« (1908/1992: 265). Anne's imagination and personality have changed her surroundings to accord with her aesthetic estimation. As in naming the nature around her, she has appropriated Green Gables into her imaginative ownership by imposing her own ideas. Place and creative mind here correspond with and reflect each other.

In addition to the links between Anne's creativity and the descriptions of nature and the house, Montgomery employs another technique to instill a sense of enchantment into her settings. Epperly has commented that »Prince Edward Island is an enchanted place where orphans suddenly find the home they have longed for and beauty and magic fairly leap from the sky and earth and sea« (1992: 31). Theodore F. Shekels makes a similar point, arguing that the island in Montgomery's novels offers a place to escape one's fears and to find acceptance and wish-fulfillment. He claims that PEI is »a place in *Anne of Green Gables* for the nurturing of Anne's being as well as the redemption of her spirit from the curse of being an orphan« (2003: 27). In Shekel's argument, the island offers an ideal place for such redemption, as it is a self-contained entity that Mont-gomery's heroine must journey to from the mainland. Part of the magic of the island is that it resembles that fairy tale land that is always ›far, far away‹. This idea of an enchanted land one can travel to, to find solace and redemption, is linked to the idealized concept of fairyland. Nicola Brown has examined fairies and fairyland during the Victorian era and argues that the Victorians found an escape from the modern world they were creating in the belief of fairyland. Fairyland was associated with the past and the pastoral, a contrast to the increasingly urban, industrialized world of the nineteenth century. Brown goes on to argue that the disappearance of fairies was more often than not associated with »ageing and the gradual drawing away of childhood« (2001: 170). Stemming from the Romantic conception of idealized childhood, maturity was considered a threat to innocence and, as in *Peter Pan* (1911), not believing in fairies, that is, growing up, would eventually kill them, at least for the non-believer.

In *Anne of Green Gables*, fairyland is the idealized domain representing Anne's imagination, which is portrayed as both Romantic and childlike. The fairyland in Avonlea is her escape from social conventions and from her orphan identity. Montgomery infuses her landscapes with elves and fairies that spring from Anne's imagination, and traces of this magic lurk in every wooded corner. Turning her surroundings into a realm reflecting her own Romantic creativity and childish fancies once more underlines the instability of reality in Montgomery's settings. PEI is represented through Anne's imaginings and consequently associated with her make-believes and dreams.

The novel opens with the famous description of the brook that runs from the Cuthbert farm past Rachel Lynde's house. Several critics have highlighted the brook's originally impulsive and uninhibited appearance, with its »dark secrets of pool and cascade« and subsequent decline, rendering it »a quiet, well-conducted little stream« (Montgomery 1908/1992: 1), and have interpreted this as the path that Anne herself must take in order to become a sensible and respected member of society. The stream, like Anne, must grow up. Rachel Lynde, who always »kept a sharp eye on everything that passed« (1908/1992: 1), represents the respectable Avonlea community and its social expectations. However, more important than the brook's conversion into a respectable stream is its origin »away back in the woods of the old Cuthbert place« (1908/1992: 1). According to the novel's narrator, nobody seems to know exactly what goes on in the Cuthbert farm woods, hence the brook's ›dark secrets‹. After all, it is only »reputed« (1908/1992: 1) that the stream was intricate and unruly at its source, beyond the reaches of society. The little stream's source, therefore, is in a space apart from society, where social rules and strict conventions do not apply. On another level, it is an extension of the escape that is the island itself and constitutes a place of mystery and adventure that is reminiscent of fairyland.

It is in the woods surrounding Green Gables, where the brook has its spring, that Anne and her friends do most of their imagining. These woods are continually infused with the presence of elves, fairies, dryads, and other inhabitants of fairyland. For example, the ring of sleigh bells is described as »elfin chimes [coming] through the frosty air« (1908/1992: 145) and there is a little pond that the girls whimsically name the Dryad's Bubble, as they imagine it to be a meeting place for nymphs. This creates an atmosphere of magic both for the inhabitants of PEI as well as the readers. The girls also build a playhouse back in the wooded area, which they call Idlewild, the name itself a reminder of the idle and untamed lives of fairies. Gammel has described this playhouse as a »hybrid space« that »blends childhood and adolescence, fantasy and reality, domestic and pagan« (2010: 238). The most enticing feature of Idlewild is the »fairy glass« (Montgomery 1908/1992: 92), which Anne imagines was lost by the fairies one night after a ball. It is this fairy space that, as Gammel argues, »nourishes alternate and independent fantasies« (2010: 238). Idlewild can be seen to represent in a small, confined space what the Prince Edward Island landscapes represent on a larger scale: a magical world, inhabited by fairies, where Anne's fantasies come to life. Idlewild is thus a sort of mirror (or rather a fairy glass) of reality in itself. The concept of mirrors and mirroring of reality is introduced early on in the novel. Anne has an imaginary friend, Katie Maurice, who lives in the reflection created by a glass door on a cupboard in her foster home. She imagines that »one day Katie Maurice would have taken [her] by the hand and led [her] out into a wonderful place, all flowers and sunshine and fairies« (Montgomery 1908/1992: 58). Montgomery here draws on a tradition of mirrors in literature reflecting reality, from *Snow White*'s mirror on the wall, to Tennyson's *Lady of Shalott* (1942) weaving her tapestry via images she sees in her looking glass.

In *Anne of Avonlea*, the idea of an enchanted realm is again picked up, indicating that Montgomery, like so many other Victorians, saw a link between childhood and fairyland. Anne, now a school teacher, and her pupil Paul Irving »both knew the way to that happy land« that »must be the gift of the good fairies at birth« (Montgomery 1909/1970: 113).

The realm of imagination in this novel, as in *Anne of Green Gables*, is referred to as a place that one can travel to. Where the Anne of *Anne of Green Gables* can literally go down to the woods and visit her make-believe play spaces, the Anne of *Anne of Avonlea* »had long ago learned that when she wandered into the realm of fancy she must go alone. The way to it was by an enchanted path where not even her dearest might follow her« (1909/1970: 19).

But then Anne and her friend Diana stumble across Echo Lodge and Miss Lavendar. The entire experience suggests a fairy tale come to life. The two girls, on their way to a friend's house, get lost and decide to ask directions at a secluded house that they pass. The scene is described as »[a]ll was very still and remote, as if the world and the cares of the world were far away,« and Anne herself hits the nail on the head, stating that she feels »as if [they] were walking through an enchanted forest« (1909/1970: 161). Echo Lodge itself is at first sight nothing out of the ordinary until Miss Lavendar introduces them to the echoes. The enchanted forest around Echo Lodge appears to be inhabited by fairies. When Miss Lavendar's maid Charlotta, described as »a messenger from pixyland« (1909/1970: 163), blows into an old horn »[t]here was a moment's stillness [...] and then from the woods over the river came a multitude of fairy echoes, sweet, elusive, silvery« (1909/1970: 166).

It is here, in this magical space, that the fairy tale romance of Anne's imagination eventually comes to life. Anne had long suspected that Miss Lavendar was not an old maid but rather an enchanted princess waiting for Prince Charming. In reality, Miss Lavendar had quarrelled with her fiancé Stephen Irving many years ago and decided to remain unmarried. Anne, who conveniently is the teacher of Mr. Irving's son Paul, decides to lend the fairy tale romance a helping hand and reintroduces the two lovers. As is appropriate for a fairy tale, the prince and the princess quickly fall in love and live happily ever after, that is, decide to marry. After the wedding, »Miss Lavendar drove away from the old life of dreams and make-believes to a fuller life of realities in the busy world beyond« (1909/1970: 238). Although Montgomery here clearly indicates that the world of reality is ›fuller‹ than the realm of enchantment left behind, there is nevertheless an undeniable suggestion that in order to live in respectable society, that is, ›the busy world beyond‹, one must leave dreams and fancies behind. It is a part of growing up and maturing; or, as Paul puts it in *Anne of the Island* when he can no longer find his Rock People: »You must pay the penalty of growing-up, Paul. You must leave fairyland behind you« (Montgomery 1915/1992: 153). Montgomery, like many Victorians, felt a deep sense of nostalgia toward the imaginative land of youth. Anne continually refuses the reality of Gilbert's love lest she too, like Miss Lavendar, must leave ›the old life of dreams‹, and the night before Diana's wedding to Fred Wright exclaims »how horrible it is that people have to grow up—and marry—and *change*« (1915/1992: 179).

The settings in the *Emily* series have a similar function to those in the Anne books. They are in themselves a version of fairyland and reflect the imagination of the protagonist. However, where the landscapes in *Anne of Green Gables* are very light and friendly, the Prince Edward Island depicted in *Emily of New Moon* is darker and more pagan. Unlike *Anne of Green Gables*, the beginning of *Emily of New Moon* is set in Emily's home, the ›house in the hollow‹. The house is first described through the eyes of society, represented by the housekeeper Ellen Greene, who thinks »it was the lonesomest place in the world« (Montgomery 1923/1993: 1). Similar to Echo Lodge in *Anne of Avonlea*, the house in the hollow is set away from other houses, secluded and mysterious, reachable by »a long, green lane« (1923/1993: 1). It looks »as if it had never been built like other houses but had grown up there like a big, brown mushroom« (1923/1993: 1). This description associates the house with nature, it is an organic form sprung from the earth, and reminds of a magical cottage dwarves or elves might live in.

It is here in the woods surrounding the little house that Emily has made her acquaintance with the magical nature of Prince Edward Island. For Emily, the landscape comes alive with fantastical creatures such as the Adam-and-Eve Tree, the Rooster Pine, and the Wind Woman. In her essay »Safe Pleasures for Girls: L. M. Montgomery's Erotic Landscapes,« Gammel argues that »the fictional island garden emerging from Montgomery's pen is a space for girls' and adolescents' emotions and desires« (2002b: 118). Where Anne and Diana can live out their eroticized friendship in Idlewild, Emily projects her eroticism onto the Wind Woman. Additionally, it is mostly nature that gives Emily her ›flash‹; that »one glorious, supreme moment« that always leaves her »tingling all over« (Montgomery 1923/1993: 7; 53). This ›flash‹ is a sort of orgasmic moment of imaginative inspiration that is directly linked to Emily's personal happiness and mental health and quite clearly a link back to the Romantic poets and their moments of sublime inspiration.

Gammel goes on to state that the Wind Woman is part of this erotic landscape as »a figure composed of Emily's idealized dead mother, mythologized nature goddess, and figure of female Eros« (2002b: 119). The Wind Woman is Emily's constant companion, comforting her and providing a sense of kinship and familiarity with the nature all around PEI. For example, after moving to New Moon, Emily's first night in her new home brings on a wave of homesickness. Unable to sleep next to Aunt Elizabeth, she lies awake, pondering her new environment until suddenly »she heard the Wind Woman at the window—she heard the little, low, whispering murmur of the June night breeze—cooing, friendly, lovesome« (Montgomery 1923/1993: 60). Emily joyfully listens to the song of her magical friend, no longer sad or afraid, and soon follows her: »[h]er soul suddenly escaped the bondage of Aunt Elizabeth's stuffy feather-bed and gloomy canopy and sealed windows. She was out in the open with the Wind Woman and the other gypsies of the night« (1923/1993: 60). It is the Wind Woman who helps Emily transcend her mortal body and fly away on the wings of her imagination, quite literally a flight of fancy.

This passage shows that the Wind Woman has another meaning than merely the eroticized incarnation of nature. The Wind Woman is a representation of Emily's imagination, much more so than the fairies in the Anne books, because the Wind Woman is part of nature itself, whereas the elves and fairies merely inhabit nature. She becomes a constant feature in the novel, recurring regularly in the nature descriptions. The Wind Woman also indicates that PEI is depicted through the eyes of the protagonist. As a reflection of Emily's imagination, the Wind Woman pervades the entire novel »providing the heartbeat of this universe« (Gammel 2002b: 119). She is not static and localized, such as the Adam-and-Eve Tree, for example, but follows Emily wherever she goes. Emily can always draw on her strength and find an escape from reality by imagining an adventure with the Wind Woman.

Houses play an important role in the settings of *Emily of New Moon*. Besides the ›house in the hollow‹, Emily forms a special relationship with New Moon, Wyther Grange, and the Disappointed House. Initially upset at having to live at New Moon, Emily soon finds that she can love her new home as much as the house in the hollow. The Murray homestead is unlike other farms in the area. Aunt Elizabeth, Aunt Laura, and Cousin Jimmy all hold with preserving Murray tradition. Hence, there are only candles, no modern dairy machines, a very old-fashioned kitchen, and various customs that to the rest of society seem hopelessly outdated. New Moon is presented as a bastion of tradition and a relic from a past era. Cousin Jimmy tells Emily that the woods surrounding New Moon are full of fairies and later on in his garden, his pride and joy, he confides that »[t]here is a spell woven round this garden« (Montgomery 1923/1993: 69). This connection of New Moon with its old-fashioned customs and its fairy-infested woods and enchanted gardens places the Murray farm on the border to fairyland as a sort of gateway or half-way house. New Moon personifies the nostalgia that Victorians associated with fairyland. Here, that the Victorians considered as the better days before industrialization are still well and truly alive and being lived. Logically, if there were a place on PEI where fairies still existed, it would be New Moon, the place that stayed in the pastoral, idealized past.

Emily is continually presented as »kin to tribes of elfland« (1923/1993: 5). She has mysterious eyes and pointed ears that hint at something not quite human, she has the ability to see the »fairy paper« in most rooms, her goal in life is to become a poetess, a profession that, according to Emily, should be »silph-like« (1923/1993: 55; 99), and she has psychic abilities allowing her to see beyond the merely factual. As an inhabitant of fairyland, New Moon seems an ideal home for Emily Byrd Starr, whose name alone indicates an other-worldly nature.

It is interesting in this context that Emily, although fully appreciative of her family's link to the past and old-fashioned traditions, is quite intent on looking to the future. She has her goal, to ›climb the Alpine Path‹, quite clearly before her and vows to do whatever it takes to reach it. The concept of time—past, present, and future—is integral to the novel and one that is closely related to space. In the woods beyond New Moon, there is a place called Lofty John's Bush, a small patch of land that once upon a time belonged to New Moon and the loss of which still smarts the proud Murrays. In this space, described as »the realm of twilight and mystery and enchantment«, a favorite playground for Emily, Ilse, and Teddy, are three paths: »the To-day Road, the Yesterday Road and the To-morrow Road« (1923/1993: 232; 129). These paths are named for their beauty with respect to time; the To-day Road is beautiful now, the Yesterday Road was beautiful long ago, and the To-morrow Road holds the promise of one day growing beautiful. The To-morrow Road also creates a sense of mystery and excitement in relation to Emily's life. It is here that Teddy whistles for Emily, a call that she feels compelled to always answer, similar to the call of the Pied Piper, and their romance unfolds.

The three paths can be read to stand in direct relation to Emily's life. The Yesterday Road is the road of her sad childhood, the time when first her mother and then her father passed away and she was left an orphan. The To-day Road is her time at New Moon, a beautiful and exciting time somewhere between her past and her future. During her adolescence she is still bound by yesterday but already able to see the potential of her future. But the To-morrow Road is obviously the most interesting of the three. Since it represents the possibilities of Emily's future, it is closely linked to her imagination, that capital with which she plans to achieve success. The To-morrow Road suggests the potential lying in Emily's creativity. It is an image that recurs throughout the three *Emily* novels and as Emily changes, so does the To-morrow Road. By the middle of *Emily's Quest*, the To-morrow Road finally belongs to Emily. Through her success as a writer, the monetary proceeds of her imaginative creativity, she is able to buy Lofty John's Bush. Symbolically, she now owns her Yesterday, her To-day, and her To-morrow in the three paths that are located in the woods.

Wyther Grange, too, is a house of importance in *Emily of New Moon*. Gammel makes a strong argument for the relationship between Wyther Grange and Priest Pond and the onset of Emily's menstruation and subsequent initiation into sexuality. Some of the indicators that she underscores are the name Priest Pond, which »signals sacred ritual«, the constant presence of water, the gulf and the many ponds in the area, which »announce the fluidity of Emily's menstruation in the landscape's imagery« (2002b: 122), and the terrifying experience in the Pink Room, an allusion to the Red Room in *Jane Eyre*, that first night. Gammel proposes that

what has been read as a simple case of nervous hysteria [which itself indicates a woman's (repressed) sexuality], or as the result of Emily's over-active imagination after having consumed too many Gothic romances, the discerning reader of Montgomery's journals will be able to decode as Montgomery's complex literary cryptogram for menstrual symptoms, depicted in Emily's bodily responses. (2002b: 123)

I would like to argue, however, that in addition to symbolizing the onset of Emily's sexual maturity, Wyther Grange also functions as the opposite of New Moon in terms of her imagination. Where New Moon is enchanted by fairies and elves, Wyther Grange, a Gothic mansion reminiscent of those in eighteenth-century novels or *Jane Eyre*'s Gateshead Hall, is haunted by »muffled sounds like tiny children's cries or moans« and »strange uncanny rustles« (Montgomery 1923/1993: 245). Where Emily can find beauty in the magical world of New Moon, she initially experiences only terror in her supernatural encounter at Wyther Grange. However, it is perhaps essential that Emily be initiated into the dark side of her own imagination, as is later proven by the paganism in the poem that Mr. Carpenter destroys in *Emily Climbs*. It is this dark imagination that will make her a better writer. Mr. Carpenter states about her pagan poem that »from the point of literature it's worth a thousand of your pretty songs« (Montgomery 1925/1993: 252). It is left unclear as to why exactly Mr. Carpenter considers it safer for Emily to stick to her less accomplished ›pretty songs‹ rather than pursue what he clearly considers more valuable from the point of view of literature. Perhaps it has to do with the fact that Emily is a woman, or that despite his cynicism, Mr. Carpenter is a good Christian. More likely, however, is that the imagination, like Christianity, seems to have a good and a bad side, symbolized by New Moon and Wyther Grange, and while both sides *can* produce literature, it is safer to stay away from the underworld of the imagination and remain in the beautiful ›secret garden‹.

Other houses of importance in the novel include the Disappointed House, which functions as a constant source of imaginative longing for Emily, the Old John House, in which Emily conceives of the idea for *The Seller of Dreams* and first realizes her profound connection with Teddy, her Aunt Ruth's house where she boards during her time at college, and the little house on the river shore that harbors the lost little boy whom Emily saves with one of her psychic visions. Although these houses play an important role in the novels, they are not as intricately linked to Emily's imaginative development as the houses discussed above and will therefore not be further discussed here.

In addition to the woods and houses connected to Emily's creativity, it is also important to note that Emily's imagination is often associated with travel or an escape from her immediate surroundings, similar to imagination in the *Anne* books. As in the example given above of Emily's first night at New Moon and her escape with the Wind Woman, many other references in *Emily of New Moon* indicate the connection between the imagination and a journey. The imagination is variously referred to as being anchored in a »secret port of dreams«, »[roaming] wonderlands of fancy«, or Emily having to write something down »before she went back from her world of dreams to the world of reality« (Montgomery 1923/1993: 124; 272; 339), to give but a few examples. This concept indicates that Emily is a kind of spirit wanderer who can escape the restrictions in her ›real‹ surroundings into a wonderful land where everything is possible. For example, during one of her erotic imaginative moments, the ›flash‹, her »soul seemed to cast aside the bonds of flesh and spring upward to the stars« (1923/1993: 80). This idea reminds readers of the Romantic poets' conception of imagination. In »Dejection: An Ode«, Samuel Taylor Coleridge writes

And oh! that even now the gust were swelling,
And the slant night shower driving loud and fast!
Those sounds which oft have raised me, whilst they awed,
And *sent my soul abroad*. (l.15–18; my emphasis)

Conversely, Emily's friendship to Dean ›Jarback‹ Priest, a man old enough to be her father who saves her from tumbling down a cliff during her visit to Wyther Grance, is both stimulating and restraining. Dean is described as an educated and charismatic man who Emily greatly admires and whose approval she seeks, particularly when it comes to her writing. He teaches Emily about books, tells her of his travels, and offers her a more alluring, almost sexually charged friendship than her relationship with boy-next-door Teddy Kent. In this new companionship with Dean, Emily »felt as if she were looking into some enchanted mirror where her own dreams and secret hopes were reflected back to her with added charm [...]. She loved him for the world he opened to her view« (Montgomery 1923/1993: 273). However, their friendship comes at a price for Emily: »Emily felt as if a cobweb fetter had been flung round her« (1923/1993: 271). She feels trapped, for she has given him power over her imagination. Because she values his opinion and desperately seeks his approval, she takes his word for truth and thus gives him authority over her own view of her creativity. She does not see that it is Dean who blocks her from her own realm of imagination, letting her only into the world of fancy he has created for her with his words and views. At one point, she comments that »[t]he wind is free—not a prisoner like me« (Montgomery 1927/1993: 87).

This ›prison‹ culminates in Emily's burning of her book *The Seller of Dreams*. Inspired by a comment made by Teddy, Emily gives herself over to her creativity and writes a wonderful, imaginative book. Feeling again the need for Dean's approval, however, she asks for his opinion. After being told that her story is »[p]retty and flimsy and ephemeral as a rose-tinted cloud. Cobwebs—only cobwebs« (Montgomery 1927/1993: 51), Emily burns her book in a rage of disappointment. It is interesting to note here that Montgomery used the image of the cobweb again, clearly indicating a connection between Emily's fetters and her creativity. Returning to the association of the wind and Emily's imagination, her tie to Dean is obviously stifling her inner vision; she is no longer free like the wind/the Wind Woman. Luckily, her love for Teddy eventually saves her from marriage to Dean and she is free once more to wander into her inner magical world at her own will.

Both the *Anne* and the *Emily* series portray a close association between setting and space and the imagination of the protagonists. These links are important in that they offer a certain view of imagination that coincides with Montgomery's own, localized creativity. Montgomery drew on her own native surroundings for inspiration and created in the Prince Edward Island of her novels a timeless land of fancy that appeals to readers of all generations the world over. Because of its links to fairyland and the enchantment conveyed in the descriptions, Montgomery's PEI displays what Fiamengo describes as »transregional magic« (2002: 232). Despite the fact that her island landscapes are firmly rooted in Canadian Prince Edward Island and display strong aspects of regionalism, Montgomery has managed to surpass the boundaries of local color and create a landscape that transcends both geographical and temporal borders. By infusing the settings with imagination and magic, the PEI of Montgomery's novels appeals to readers of all nations and ages. The settings have taken on an aspect of timelessness and universality despite its localized heritage.

In terms of imagination and childhood/adolescence, both Montgomery's heroines outgrow—at least in part—their youthful make-believes. Anne, like Miss Lavendar, eventually marries and gives up her youthful scribbling in order to join the adult world of reality, that is, she becomes a housewife and mother, never writing the thrilling novel or inspired poem she once dreamed of. Anne, like so many other literary heroines, falls victim to the appropriate stereotyped happy ending and lives happily ever after. *Anne of Ingleside* (1939), the last novel in the Anne series that focuses on Anne (the last two books largely deal with Anne's children) ends with a very poignant scene. Anne is about to go to bed after a dinner party:

In her white gown, with her hair in its two long braids, she looked like the Anne of Green Gables days... of Redmond days... of the House of Dreams days. That inward glow was still shining through her. Through the open doorway came the soft sound of children breathing. [...] Anne grinned. [...] »What a family!« Anne repeated exultantly. (Montgomery 1939/1983: 263-264)

Although she still looks like that young orphan girl who came to Avonlea many years ago, the Anne depicted here is a completely different person. She is no longer a child, no longer an orphan, and no longer unloved and unwanted. Despite the comment that she has kept her inward glow, the implication is that this glow is now grounded in her new reality as wife and mother, rather than her childhood make-believes. A few pages earlier she explains her failed writing career: »I'm writing living epistles now« (Montgomery 1939/1983: 254). In other words, Anne sees herself as author of her children's lives.

Emily is left with a slightly more ambiguous ending. Although she, too, grows up and finds love with Teddy Kent, both she and Teddy are artists and it is implied that they will continue to follow their respective arts together. *Emily's Quest* ends with a metaphoric joining of Emily and the Disappointed House: »She was standing where the To-morrow Road opened out on the Blair Water valley. Behind her she heard Teddy's eager footsteps coming to *her*. Before her on the dark hill, against the sunset, was the little beloved grey house that was to be disappointed no longer« (Montgomery 1927/1993: 228). By using both the To-morrow Road and the image of the house, Montgomery implies that Emily's future adult life will be a happy one, that is, no longer disappointed. That happiness, however, seems to come from Teddy, whose footsteps are heard in the distance, and their little grey house, altogether a very domestic image despite the atmospheric, magical sunset that once gave Emily her ›flash‹.

In conclusion, it seems that although Montgomery herself was a very progressive woman who became a best-selling author, her heroines did not follow in her footsteps. Neither Anne nor Emily managed to maintain or develop their youthful imagination in adulthood. For both girls, Prince Edward Island functions as an imaginative playground during their childhood, allowing them the freedom to expand their minds only to return to reality as grown women.

References

Brown, Nicola (2001): *Fairies in Nineteenth-Century Art and Literature*. Cambridge: Cambridge University Press.
Chopin, Kate (1899/1993): *The Awakening*. New York: Dover Publications.
Coleridge, Samuel Taylor (1817/1983): *The Collected Works of Samuel Taylor Coleridge: Biographia Literaria, Volume I*. London: Routledge.
Coleride, Samuel Taylor (1802/1996): »Dejection: An Ode«, in Margeret Ferguson/Mary Jo Salter/Jon Stallworthy (eds.), *The Norton Anthology of Poetry*, 4th Edition. New York: Norton, pp. 760–763.
Epperly, Elizabeth Rollins (1992): *The Fragrance of Sweet-Grass: L. M. Montgomery's Heroines and the Pursuit of Romance*. Toronto: University of Toronto Press.

Fiamenga, Janice (2002): »Towards a Theory of the Popular Landscape in Anne of Green Gables«, in Irene Gammel (ed.), *Making Avonlea: L. M. Montgomery and Popular Culture*. Toronto: University of Toronto Press, pp. 225–237.

Gammel, Irene (2002a): »Introduction«, in Irene Gammel (ed.), *Making Avonlea: L. M. Montgomery and Popular Culture*. Toronto: University of Toronto Press, pp. 3–16.

—— (2002b): »Safe Pleasure for Girls: L. M. Montgomery's Erotic Landscapes«, in Irene Gammel (ed.), *Making Avonlea: L. M. Montgomery and Popular Culture*. Toronto: University of Toronto Press, pp. 114–130.

—— (2010): »Embodied Landscape Aesthetics in *Anne of Green Gables*«, *The Lion and the Unicorn* 34 (2), pp. 228–247.

Mallan, Kerry (2003): »Secret Spaces: Creating an Aesthetic of Imaginative Play in Australian Picture Books«, *The Lion and the Unicorn* 27 (2), pp. 167–184.

Miller, Kathleen Ann (2010): »Haunted Heroines: The Gothic Imagination and the Female *Bildungsromane* of Jane Austen, Charlotte Brontë, and L. M. Montgomery«, *The Lion and the Unicorn* 34 (2), pp. 125–147.

Montgomery, L. M. (1908/1992): *Anne of Green Gables*. New York: Bantam Books.

—— (1909/1970); *Anne of Avonlea*. London: Penguin Books.

—— (1915/1992): *Anne of the Island*. New York: Bantam Books.

—— (1939/1983): *Anne of Inglside*. London: Penguin Books.

—— (1923/1993): *Emily of New Moon*. New York: Dell Laurel-Leaf.

—— (1925/1993):*Emily Climbs*. New York: Bantam Books.

—— (1927/1993): *Emily's Quest*. New York: Dell Laurel-Leaf.

Shekels, Theodore F. (2003): *The Island Motif in the Fiction of L. M. Montgomery, Margaret Laurence, Margaret Atwood, and Other Canadian Women Novelists*. New York: Peter Lang.

Shelley, Percy Bysshe (1840/2009): »A Defence of Poetry«, in *English Essays: From Sir Philip Sidney to Macaulay*. New York: Cosimo Books, pp. 345–380.

Smith, Elsie L. (2008): »Centering the Home-Garden: The Arbor, Wall, and Gate in Moral Tales for Children«, *Children's Literature* 36, pp. 24–48.

Solt, Marilyn (1984): »The Uses of Setting in *Anne of Green Gables*«, *Children's Literature Association Quarterly* 9 (4), pp. 179–198.

Fallujah Manhattan Transfer
The Sectarian Dystopia of *DMZ*

GEORG DRENNIG

The comic series *DMZ* (2005–2012), written by Brian Wood and mainly illustrated by artist Ricardo Burchielli, repeatedly confronts the reader with familiar images—burned out military vehicles and aircraft, and U.S. Army soldiers with iPods strapped to their helmets fighting against a hard-to-identify enemy that hides among the civilian populace. Though numerous items of the setting bear a deliberate resemblance to images from the U.S. wars in Iraq and Afghanistan, the fighting in *DMZ* takes place in Manhattan, the key symbolic battleground of the series' dystopian vision of a United States embroiled in a civil war of unclear frontlines. This is, however, not a mere change of setting for a scenario of asymmetric warfare and insurgency, the machinations of the military-industrial complex, or sectarian infighting on city streets—though all of these are part of *DMZ*—but a commentary on the fractured state of U.S. society itself and its tribalism. It is a dystopia based on the failure of the utopian project known as United States of America and the numerous smaller heterotopian spaces contained within the comic series itself.

Heterotopias, first conceptualized by Michel Foucault, are »places that do exist and that are formed in the very founding of society—which are something like counter-sites, a kind of effectively enacted utopia in which the real sites, all the other real sites that can be found within the culture, are simultaneously represented, contested, and inverted« (1986: 24). Foucault formulates several categories of heterotopias, all of which share the quality of being spaces in which societal rules and norms are renegotiated or contested. His ruminations on the topic include a variety of sites that can be so considered and offer an equally broad range of principles that may apply to them. These principles, however, do not provide a strict classification or a determined set of criteria, thus allowing a wide-ranged conceptualization of ›other places‹. This conceptual ambiguity and openness to interpretation is, in fact, a quality of the heterotopia itself:

it is a space of »experimentation, fluidity, and disorder« (Mitchell 2000: 215), emphasizing the »possibility of possibilities« (Reid-Pharr 1994: 348). It is an impermanent space of alternatives to the strategic system, yet provided and even controlled by it.[1]

The utopian dimension of heterotopias becomes more striking in *DMZ*, since it contrasts with the series' general dystopian vision. ›Dystopia‹ denotes a fictional society that represents not simply a worst-case scenario forecast for the future, as the term is usually defined,[2] but adds a critical component to it: it is a forecast grounded in the time of writing and relying on the readers' familiarity with the fault lines in contemporary society. In the series, Wood provides numerous references to trends and events familiar to a North American audience, thus inextricably connecting the story's dystopian setting to contemporary U.S. society.

One of the first images in the comic establishes this basic setting by showing a map of New York City, with »American troops dug in on [the] coastline« (Wood 2005: 7) of Long Island, the insurgent Free States controlling New Jersey, and Manhattan as the De-Militarized Zone that gives the series its title.[3] This division is the result of the rebellion of the Free State Armies, an uprising of »Middle America [...] out of frustration, anger, and poverty to challenge the government's position of preemptive war and police action throughout the world« (Wood qtd. in Richards 2005: par. 3). The civil war that ensues has been going on for years in the series' undated timeline, »fought in bits and pieces all over the country,« without clear »borders of front lines« (Wood 2007: 65). Throughout the entire comic, Wood avoids any clear partisan designations—Republicans and Democrats are not mentioned once—stating in an interview that »the two warring groups in ›DMZ‹ are just extremists fighting extremists« (qtd. in Richards 2005: par. 4). This vague scenario becomes more concrete, however, in a key symbolic space of American culture: Manhattan; though, as will become evident below, the borough is not a uniformly important space itself.

The island of Manhattan is »the one bit of country that neither side could manage to claim« (Wood 2007: 141), yet it features as the site of the civil war's major symbolic struggle. The efforts of the Free States of America and the remaining U.S. government to expand their footholds there and to control the narrative of what the DMZ is—haven for insurgents or enemy troops and victim of the other side's actions—not only propel the comic's plot forward, but point to the importance of Manhattan as a symbolic space. Ground Zero, and control over it, is accorded a special role in this respect, and several storylines within the series mention the U.S. government's struggle to retain the ability to use it as a propaganda tool. Wood stated in an interview »that *DMZ* would not exist if 9/11 and the Bush administration hadn't found love with each other«. And while he went on to explain how he wasn't sure if the »book need[ed] them anymore to

Illustration 1: Map of the DMZ. Note the focus on Lower Manhattan and Ground Zero. The inserts in the white fields represent broadcasts by Liberty News. The provenance of the map within the comic is not stated.

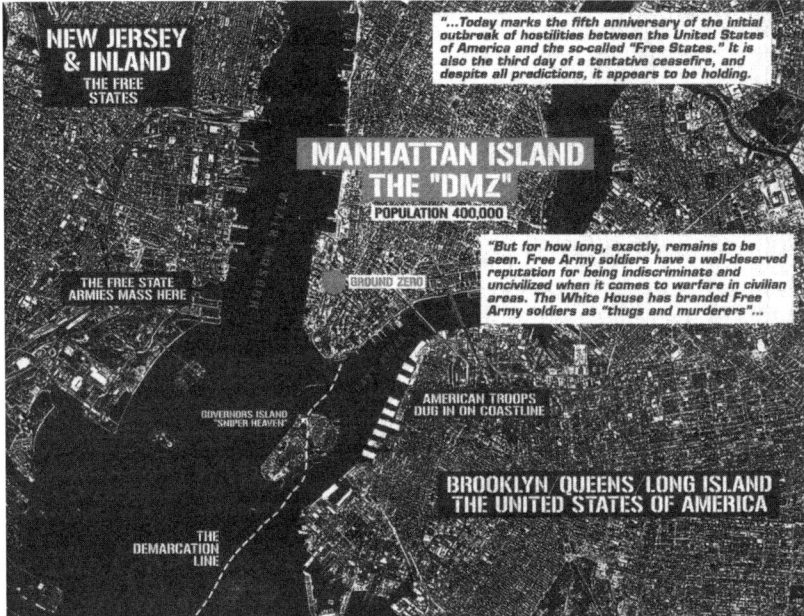

Source: Brian Wood, *DMZ: Body of a Journalist*, p. 7. *DMZ: Body of a Journalist* © DC Comics, 2007.

stay topical« (qtd. in Jaffe 2009: par. 20), the early issues especially rely on the readers' familiarity with the foreign and security policy of the George W. Bush administration.

At first sight, then, *DMZ* seems like a meditation on the War on Terror, the invasion of Iraq, and also on the media landscape of the early George W. Bush era. One of the key actors throughout the series is Liberty News, a media conglomerate completely in line with the remaining U.S. government, repeatedly referring to itself as ›News for America, and Americans‹. As a page of exposition in *DMZ*'s volume four points out, »What is left of the United States has merged with its biggest supporter in the media, Liberty News, to the point that the two are indistinguishable« (Wood 2008a: 8). Tellingly, Army soldiers give clear directions to journalists on how to frame their killing of a family of four: »This is: ›Insurgent Cell defeated en route to engage American forces‹ or something. Whatever. And crop out the small bodies« (Wood 2006: 64). While this general situation of complete media–state collusion is clearly dystopian amplification, the depiction of Liberty News' unquestioning propaganda, especially regarding events on the battlefield, forges a strong reference to U.S. media behavior during battlefield operations in Iraq.

Illustration 2: U.S. troops engaged in unspecified conflicts—though the civilians and enemies change, the pictorial representation of warfare does not.

Source: Brian Wood, *DMZ: Friendly Fire*, p. 91. *DMZ: Friendly Fire* © DC Comics, 2008.

The comic's text is even more explicit in drawing parallels when it refers to another organization that remains essential to events throughout the series, Trustwell Inc., »a reconstruction/security firm, equal parts Halliburton and Blackwater« (Wood 2008a: 8). *DMZ*'s heavy use of images that have attained iconic status encourages the drawing of parallels to the real world: throughout the series, the reader encounters hooded prisoners in orange jumpsuits, scenes of torture, and U.S. Humvees wrecked by improvised explosive devices. The detention facility at Guantanamo, the abuse at Abu Ghraib, and the violence of the Sunni insurgency thus loom large—not only in the background, but in the comic's panels. Therefore, when the protagonist asks himself the question whether »sending roving packs of young soldiers out into a civilian area with shitty training and no intel and expecting results [is] a defensible act« (Wood 2008a: 116), the ›civilian area‹ in question is as much Manhattan as it is Iraq and Afghanistan.

Reading *DMZ* as a commentary on U.S. action abroad would, however, neglect a deeper dystopian dimension of the series. When the protagonist, journalist intern Matty Roth, enters Manhattan as part of a Liberty News crew and finds himself stranded after the army helicopter that took him in is shot down, the microcosm he explores is symptomatic of a larger American malaise. Wood's DMZ and those fighting over it are symbolic of a fragmented American society increasingly marked by groups that only achieve further division in asserting their interpretation of America—itself a utopian project—to be the correct one. Matty Roth's experiences of quasi-sectarian warfare in Manhattan point out the lines dividing contemporary U.S. society: ethnic, ideological, and spatio-economic fault lines which, in a dystopian worst-case scenario, lead to a New York City that is uncannily similar to Fallujah at the height of the Iraqi Sunni insurgency.

When he first enters Manhattan—described by the series' author as »equal parts *Escape from New York*, Falluja, and New Orleans right after Katrina« (Richards 2005: par. 9)—and thus initiates the series' plot, Matty Roth is part of the problem. Having grown up on Long Island during the early years of the civil war between the Free States and the U.S. and working for the propagandistic media outlet Liberty News, his knowledge of both the DMZ and the FSA reflects his extremely limited news diet, a fact he realizes early on when he justifies his behavior: »You don't understand what we're told over there. I didn't know so many civilians still lived here. All we hear is about insurgents and stuff« (Wood 2006: 26). He then struggles to find a journalistic voice of his own within the DMZ, becoming a freelance reporter working in fleeting arrangements for both Liberty News and its international competitors. Through both his narration and occasional inserts of Liberty News reports that stand in jarring contrast to what the reader has just been presented in the comic's pages, the problem of media segmentation is repeatedly thematized. Though the exact media landscape of the war-torn country is never explained, it becomes clear that reporting from and about the war exacerbates the existing divisions. In Wood's dystopia, the fourth estate is yet another dividing force.

Yet the ultrafragmentation of society in *DMZ*'s Manhattan results not only from larger power struggles taking place across the continent—in an aside, Wood plays with the idea of Quebec having become an independent force (2006: 125)—but of previously existing fault lines that mirror those of contemporary American society. Five years after the Free States have declared war on the central government, »Manhattan's a fucked up place,« since »the ›rules‹ change from block to block, neighborhood to neighborhood« (Wood 2006: 12), as a U.S. Army soldier explains to the yet-uninitiated Matty Roth. The ethnic and ideological militia of the city have created a DMZ that »is all about enclaves, zones of specific influence that sometimes are defined by neighborhood lines but more often than not have their own rules and borders« (Wood 2011: 35). These rules and borders, however, are not per se caused by the civil war: they reflect preexisting divisions within the national and the local society. The dystopian strategy of *DMZ* is thus to take social fissures that readers can recognize from U.S. society and then spatialize and amplify them to the point where the resulting scenario bears an uncanny resemblance to the media portrayal of Iraq beyond the aforementioned iconic level: Manhattan becomes a wrecked city torn apart by sectarian infighting.

As mentioned above, the divisions that form the basis for this transfer of Fallujah to Manhattan are primarily ethnic and ideological: When Matty Roth first lands in the DMZ, he finds himself in territory held by radical Trotskyites. In contrast to such leftist groups, several of which are shown to exist in the series, one of the few militia that receives a more detailed portrayal in the series, the right-wing Nation of Fearghus, is a group that Matty has heard to be »White Power, some say« (Wood 2008a: 70). Yet class, already a border-creating force in pre-DMZ Manhattan, is another factor that is explored in its full divisive logic. Armed inhabitants of the Upper Eastside raid Central Park for provisions, because they are not even attempting to make arrangements with others. Artist collectives, too, stake out their territory through violent means. Most of these groups, however, share a claim to New York or neighborhood allegiance and authenticity. As the leader of the Nation of Fearghus states: »We're New Yorkers, born and bred. Defending our home. We're also not traitors. We'd be U.S. military if those fucks cared about this city« (2008a: 70). ›Home‹, however, becomes a term fraught with violence in the dystopian setting—in *DMZ*, pre-war borderlines of identity politics are policed with automatic weapons.

Only three groups in the series defy this dynamic of societal divisions taken to the extreme that is at the core of Wood's dystopian vision. For the environmentalist militia known as Ghosts of Central Park, New York holds a pastoral promise of sustainable urbanity. Although one can read them as yet another ideology-based group asserting their own territory, their—ultimately unsuccessful—agenda is primarily one of securing a better future for the city after the end of hostilities. Wilson, former triad gangster and ›guardian‹ of Chinatown, and one of the series' major actors, is equally interested in preparing for the post-war situation. Though his power base, rhetoric, and self-fashioning seem to make him yet another ethnic leader, the comic slowly reveals his intentions of ending up as ruler over entire Manhattan. His warning to those outside of Chinatown, »When you fight for New York, don't fight in our name. You don't represent us« (Wood 2008b: 63), is thus not just a rejection of other groups' claims of an allegiance to their city, but also an assertion of his status as more than the leader of another ethnic militia. This special positioning is also part of a long-term strategy that assumes New York will emerge from the final dissolution of the Union as an independent state.

The forces based in the Empire State Building then provide something of a polar opposite to the Ghosts and Wilson. Consisting of former first responders who have lost their families at the onset of the war and during the incomplete evacuation of Manhattan, they pursue a nihilistic goal: »Seek out signs of civilization, signs of military presence, [...] signs of life, of anything growing like a mushroom on this shitheap of a city. And kill it« (Wood 2010: 16). In order to create a sense of unity and determination that even makes them willing to become suicide bombers, their leaders put them through daily sessions of remembering and grieving—a thinly veiled authorial comment on the utility of traumatic memory for violent purposes. Yet Wood links their actions to other events in the series and hints at some hidden agenda of their leaders, ultimately leaving their precise role or possible allegiance in the larger conflict unclear.

This unclear and ambiguous nature of the majority of agents in the conflict is a major feature of the series. In reference to the non-alignment of the FSA and the U.S. to partisan distinctions between Republicans and Democrats, Wood stated in an interview that »the politics of such a conflict can be a little weird« (Richards 2005: par. 4). Events after the launch of the series in 2005 have changed potential readings of the conflict's political poles, for the posturing of the Free States' rebellion against the central government does resemble Tea Party rhetoric, and though the Gadsden flag that appears in a piece of exposition late in the series is not explicitly linked to the Free States, it is clearly a part of the symbolic representation of crisis and disunity. On the other side of the political spectrum, Barack Obama's foreign policy has often been criticized for being insufficiently different from that of his predecessor. Still, both the visual language of *DMZ* and the statements of characters serve to discourage simplistic partisan designations.

The anti-federal rebels that rise in the Midwest are portrayed in a wide variety of visual styles,[4] sometimes in line with the ›redneck‹ moniker assigned to them by U.S. forces in *DMZ*, some even proudly sporting neo-Nazi paraphernalia. Other FSA fighters, including one of their key representatives in the plot,

Illustration 3a and 3b: The visual representation of the FSA Lincoln Tunnel Commander and the unnamed black FSA soldier runs counter to their right-wing/redneck orientation.

Sources: Brian Wood, *DMZ: Body of a Journalist*, p. 23. DMZ: Body of a Journalist © DC Comics, 2007 (left image). Brian Wood, *DMZ: War Powers*, p. 20. DMZ: War Powers © DC Comics, 2009 (right image).

are, however, drawn in a way that is inconsistent with visual codes for right-wing/›redneck‹ orientation: the political stance that would usually be implied by the FSA commander's[5] long hair or the unnamed black soldier's dreadlocks is at odds with the image of the Free States as a conservative, right-wing, possibly Christian fundamentalist project (see *Illustration 3*).

Statements made to Matty Roth by the Free State Armies' supreme commander that his men »play up the redneck angle at times« or that Miami is »too conservative« (Wood 2011: 25–26) for his forces to hold create a further sense of ambiguity, similar to the way that the extremely hawkish and conservative rhetoric of U.S. politicians and Liberty News discourages a simple reading of the remnant U.S. being the ›liberal‹ side in the conflict. In the dystopian American society of *DMZ*, such distinctions have given way to different, irreconcilable and unable-to-democratically-coexist positions; the »extremists fighting extremists« (Richards 2005: par. 4). Yet, just as the militia does in Manhattan, both sides lay claim to representing the same thing: theirs, too, is a struggle over being the true Americans in the civil war.

DMZ's U.S. borrows from historical precedent in their rhetoric of ›America‹. Thus, a speech of the President—whose party affiliation remains unnamed throughout the series—refers directly to the War of Secession: »There will be no dissolution of this great union of ours [...]. We've gone to war to preserve the dream of our founding fathers, and we'll go to war again« (Wood 2007: 133). This assertion and the general U.S. rhetoric, repeated in different phrasings by its soldiers and other agents throughout the series, not only accuses the Free States of being traitors to the administration, but to the American idea itself: »America, man. Love it or leave it« (Wood 2009b: 35), as a U.S. army soldier tells his FSA captive at gunpoint.

The Free States reject such a construction, and when Matty Roth meets the Holland Tunnel commander—the main FSA representative in the series—for the first time, said commander greets him with, »You're looking at the truest, most bluest motherfucking Americans you'll ever meet« (Wood 2006: 125). When they meet again on the Fourth of July—still celebrated in the DMZ—the commander tells him: »We'll make an American out of you, yet« (Wood 2007: 50). As the Free States' rising is portrayed as an extremely loose coalition of interests, primarily united by resistance against the country's foreign policy and the handling of the economic crisis, and the political lines in the conflict are blurry, a reading of the FSA as a secessionist force is discouraged.

This is made especially clear when, in their attempt to show their ›truly American‹ credentials, both sides in the war agree on holding elections for the DMZ. Matty Roth's comment that »as fucked up and fractured as this country is, it's still not so far gone that it's giving up on the notion it's a democracy« (Wood 2009a: 17) implies a unity of purpose, or rather, a unity in laying claim to representing the ›true‹, or ›real‹, America. It also acknowledges the fundamentally utopian aspiration behind the founding of the United States. Yet it is exactly this claim of being heir to the founders' project that has made the war possible to begin with. As is the case on the streets of *DMZ*'s Manhattan, the dystopian vision of the comic relies on an uncanny resemblance to the current state of the Union, in which the centrifugal forces cause discord by laying claim to a central idea and identity. This claim, in turn, is not exclusively symbolic: the two sides' struggle to physically control one of the key symbolic spaces of the country is the major force driving the plot.

Matty Roth, however, soon realizes that although it is highly valued as a symbolic space by both sides, the reality of life in anything-but-demilitarized Manhattan is underrepresented. The Manhattan he explores and tries to become a media voice of is a laboratory of urbanity and a hotbed of cultural activity, albeit a highly dangerous one. The same New York pride the diverse militia in the DMZ articulate—with the aforementioned exceptions—fosters the emergence of a spatial alternative of sorts. As Matty Roth states, »this is a city stuck between everything [...]. Certainly doesn't feel like America [...]. This is a whole new tribe, a new culture« (Wood 2007: 114). He thus asserts the emergence of Manhattan as a heterotopia within the setting of the civil war: an ambivalent space of real, lived, existences not accounted for by the strategic narratives of the two warring national factions, a potential, tactically articulated, alternative to the strategically determined dichotomy of war.⁶ In other words, the culturally and politically new aspect of Manhattan and its in-between-ness make it a spatial alternative that, while marginalized from the viewpoint of power and representation, carries significant potential in offering an escape from siding with either the Free States of America or the United States.

Several issues of *DMZ* are dedicated to this ›new culture‹ in the marginalized space and the artists articulating the underrepresented experience the city's resilience and life during wartime. The comic also features short pieces that function as a travel/club guide compiled by the narrator, artist profiles, and full stories exploring the intersection between an organic arts scene in the DMZ and the civil war and global culture. In »Random Fire« (Wood 2008b: 97–119), one such story dedicated to an artist, a musician and producer of worldwide renown is brought into Manhattan to bolster his credibility as ›authentic‹ and ›edgy‹. Yet the plan to stage a shootout from which the international guest, DJ Grendel, is to emerge unscathed but with enhanced ›street cred‹—a plan that involves Trustwell Inc. mercenaries—is foiled by Random Fire, a Manhattan hip hop artist choosing ›authenticity‹ and allegiance to the DMZ over the commercial gain that would be involved for him. The importance of artists for the city is duly noted by Matty Roth, who considers them »custodians of the soul of New York City, embodiments of hope against the forces arrayed against us« (Wood 2011: 36).

The local pride and hopefulness expressed by the DMZ's artists—some of whom even form their own militia and establish zones of influence—can also be read as an expression of a more basic form of self-reliance. With Manhattan being mostly cut off from supplies of all kinds, its inhabitants are forced to find ways of sustaining themselves, which results in practices that bear resemblance to contemporary redefinitions of cities as laboratories for a more sustainable future. This clearly references trends such as urban or guerilla gardening, or locavorism, and ties in with discourses of sustainability within the city. When Matty Roth's guide Zee explains that an urban setting is conducive to growing food, and that she has »friends that have these awesome rooftop greengardens, and others that grow sprouts and make tofu and falafel in their basements« (Wood 2006: 37), she is, however, not talking about a cultural practice in search of a lost authenticity or way of subverting global food markets. As Sarah Jaffe has observed during an interview with Wood, the »references to sustainable living that the residents were forced into« are present »not out of high-minded ›green‹ concerns but because it was how they had to live« (2009: par. 28). The potential of the DMZ as an urban laboratory and the emergence of it as a heterotopian alternative to war-torn America are thus not deliberate attempts to pioneer new modes of urban living, but the ways that a besieged city finds to survive.

In looking-distance to Manhattan, the protagonist visits another marginal location that sees the emergence of a spatial alternative to the two warring factions' division of the country. On U.S.-held Staten Island, Matty Roth encounters armed forces making use of the fact that their high-level commanders have directed their attention elsewhere. The benefit of guarding the ›forgotten borough‹ instead of a symbolically more valuable area allows the troops to engage in partying, drug consumption, and the full use of the operational leeway afforded by their location. The non-importance of the site even allows for fraternization between the soldiers of both armies: Matty Roth witnesses a visit by FSA forces who then proceed to party with their U.S. counterparts. This arrangement, however, does not, or, rather, cannot, last in the dystopian comic series. After a vial of a chemical warfare agent goes missing, the U.S. officer in charge takes his FSA guests as prisoners. The heterotopia of *DMZ*'s Staten Island, with its potential for reconciliation, gives way to uncannily familiar images of American soldiers guarding and torturing hooded prisoners in orange jumpsuits.

Staten Island is not the only marginal and overlooked space that becomes significant. The elections that massively change the power dynamics in the DMZ are won by Parco Delgado, a candidate from Upper Manhattan, another symbolically-less-important space. This surprise works on several levels, as it uses both readers' and fictional characters' Mid- and Downtown-focused mappings of Manhattan. Even *DMZ*'s protagonist at first fails to understand the situation. An early mapping he creates reflects his bias: »The people are all downtown, as is most of the food and power and culture. Uptown is the realm of animals, shut-ins, looters, and mass graves« (Wood 2007: 152). A monologue of another character, however, reveals both the falseness and utility of this misconception and highlights the symbolic importance of different parts of Manhattan: »A lot of us headed uptown. Little colonies formed in the old projects, away from the hot zones. The war was far to the south, and while we suffered from it in some ways, at least we didn't have bombs coming down on us while we slept« (Wood 2008b: 43). Safety in the second civil war is thus a matter of being in a little-mapped space. As shown by the uprising of the Free States starting in ›flyover country‹ or by the—fragile and ultimately unsustainable—ceasefire on little-noticed Staten Island, such spaces hold a dynamic potential. The electoral upset by Parco Delgado, one of the most massive changes in *DMZ*'s political situation, in fact, is only possible because upper Manhattan is paid so little attention to by the U.S.: For a brief moment in the series, it seems as if New York's »›forgotten population‹ will determine the future of this city« (Wood 2009a: 136), that is, »the ones not in the foreground, the ones not running militia or working with Trustwell or running sustainable local businesses or pirate radio« (Wood 2009b: 76).

Though the rise of Parco Delgado in DMZ politics later turns out to have been facilitated by the Free States, the rhetoric and strategy of the Delgado Na-

Illustration 4: Parco Delgado and his bodyguards

Source: Brian Wood, *DMZ: Blood in the Game*, p. 19. *DMZ: Blood in the Game* © DC Comics, 2009.

tion is that of a revolt of those marginalized by the civil war's two major factions and the relatively privileged inhabitants of Lower Manhattan. The movement appropriates twentieth-century revolutionary iconography: Delgado's troopers wear red berets with single stars, reminiscent of the iconic Che Guevara picture (see *Illustration 4*), and, in contrast to the differently equipped FSA and U.S., often carry AK-47s. When Parco Delgado dismisses both the FSA's and the U.S.' claim to representing the real America, he articulates this rejection as giving voice to an unheard group: »How do you know what is real? How can you speak for so many people whose options are so limited?« (Wood 2009a: 19). Though Matty Roth is highly impressed by this rhetoric, eventually becoming a supporter and collaborator of Delgado, other characters take a less appreciative stance, characterize the populist leader as »the bastard child of Hugo Chavez and Al Sharpton« (2009a: 69), and deconstruct his posture: »[H]e's not some romantic freedom fighter for the people. He's not Che, he's not Mao, he's not Chavez, nothing like that« (2009a: 48).

The electoral victory of the movement, however, turns out to bring only ephemeral gains. Matty Roth's newfound power as the Delgado administration's public voice only leads to hubristic behavior on his behalf, alienation from his friends and allies, and with him ending up as an »enclave of one« (Wood 2009b: 141), while on the larger political scale, the new ruling faction is unable to consolidate its power sustainably. Again, it becomes clear that the dynamic potential of marginal spaces is severely limited in the strategic dystopian setting of *DMZ*. In the case of Parco Delgado, it is, indeed, his attempt to leave the marginal position in the larger power struggle that leads to yet another drastic shift in the series' plot and tone. With the help of the FSA, the Ghosts of Central Park—themselves trying to consolidate their gains in creating an environmentalist alternative of their own—and Matty Roth, Parco Delgado obtains a nuclear warhead. The public statement to the international press following this acquisition asserts Manhattan as a strategically independent entity: »The City of

Manhattan is now a nuclear-armed state [...]. The people of this beleaguered city voted and gave us independence, an identity, a nation. We hope the world realizes that we are prepared to deter and defend that at all costs« (Wood 2010: 80–81). This also means abandoning the status of an in-between heterotopian space, as the Delgado Nation becomes a determined strategic actor of its own.

Yet this play for power initiates the penultimate act in the civil war and the fighting over the DMZ, as it provides the U.S. with a pretext to use the full force and technological advantage that it holds at its disposal. The reaction to, and search for, Parco Delgado's weapon of mass destruction becomes, as a radio voice in the comic states, a meta-commentary »on the nature of preemption while evoking the legendary ›missing WMDs‹ of wars gone by« (Wood 2010: 121). When U.S. forces detect the location of the warhead hidden inside their own territory bordering on New York City, they initiate a false-flag operation: Detonating a tactical nuclear weapon of their own and assigning the blame to the Delgado Nation, the U.S. creates a pretext for a massive military response. At the end of volume eight of the series, the experiment of a nation of New York has failed: a bruised and battered Matty Roth, who has just experienced a tragic personal failure of his own ambitions to become a powerful political agent in the DMZ, gazes north along a Manhattan avenue and sees a mushroom cloud rising in the distance. In the massive airstrikes that follow as ›retribution‹ and preparation for the invasion of ground forces, any remaining alternative spaces—Chinatown and Central Park, but also Manhattan itself—are obliterated or subdued by U.S. military. The protagonist, remembering his role as witness and acting accordingly, realizes the futility of attempts to create a heterotopia in *DMZ*'s dystopian strategic setting: »Most of my time here has been spent trying to prove how alive the city is, despite it all. The undefeatable spirit of the people, against all the odds. But this is a dead city« (Wood 2011: 57).

Following the nuclear explosion, the U.S. has world opinion on its side, and its troops return to Manhattan, defeating the Free States forces and ending the war in an armistice. During his last days in the city—he is eventually arrested, tried for treason, and imprisoned for life—Matty Roth explores the emerging post-war order in Manhattan, and finds that the fault lines that split the DMZ, as much as American society at large, have won out against the elusive heterotopias that the war had briefly made possible.

Though the second civil war has thus ended, the core issues that make Wood's vision dystopian—the divisive potential of the fault lines running through American society and the contestation of the meaning of ›America‹—remain, both in the fictional world of *DMZ* and outside. The insurgency of the Free States is only a symptom of a larger malaise, as Wood acknowledged in a comment on the relation of *DMZ* to the Bush era: »Will we be able to get back to the pre-Bush status in this country? And was that really all that great, back then« (qtd. in Jaffe 2009: par. 19)? Denying the other camps' interpretation of what America means and laying claim on that central term—from whichever perspective—is not bound to the foreign policy exploits of a particular administration. Nor is the strength of intra-societal boundaries, whether they are ethnic or socio-economic. The hopeful potential of alternative spaces within this dire strategic setting of a nation divided remains unrealized in *DMZ*. The heterotopias explored in the series—whether they are a Manhattan identity creating ›something new‹ or the temporary fraternization between opposing armies on Staten Island—are all exposed as unsustainable and subject to tragic dissolution. What remains is sectarian warfare, transferred across cultures and continents from Fallujah to the streets of Manhattan.

Notes

1 | For further discussion that has informed the usage here, see Soja (1989: 16-21). The considerations in this chapter—regarding both Foucault's concept and the larger matter of power and space—also borrow especially from Don Mitchell's *Cultural Geography*.

2 | Compare, for example, with the definition in John A. Cuddon's *Dictionary of Literary Terms and Literary Theory*; there, dystopia is defined as the converse of utopia, which, in turn, refers to a wide range of divergent visions of a perfect society (1991: 1016-1019).

3 | For an analysis of the way that the grammar and language of comics are used to construct Manhattan and make sense of the physical setting in *DMZ*, see Sanchez-del-Valle (2011).

4 | *DMZ*'s serial mode of production over several years and the occasionally changing artists collaborating with the writer clearly work against consistency in the visual portrayal of secondary characters.

5 | Readers are given some insight into the commander's views on isolationism as a desirable foreign policy in *Free States Rising*, the eleventh volume of the series; see especially Wood (2012: 14-25).

6 | Both the concepts of the ›tactical‹ and the ›strategic‹ are used in the sense of Michel de Certeau (1980/1988: 23-24), that is, as concepts of space and power in which the strategic is the organized, institutional, large-scale determinant within which tactics are the methods used by those relegated to a weaker status to make and leave their own meanings in the larger strategic space. From a de Certeauvian point of view, a heterotopia is thus by definition a place of tactical alternatives to the strategic setting.

References

Cuddon, John A. (1976/1991): *Dictionary of Literary Terms and Literary Theory*, 3rd Edition. London: Penguin Books.

de Certeau, Michel (1980/1988): *Kunst des Handelns* (trans. Roland Voullié). Berlin: Merve Verlag.

Foucault, Michel (1986): »Of Other Spaces« (trans. Jay Miskowiec), *Diacritics* 16, pp. 22–27.

Jaffe, Sarah (2009): »Brian Wood: State of the *DMZ*«, *Newsarama* [online], 16 March, http://www.newsarama.com/comics/030916-Wood-DMZ.html. 2 December 2011.

Mitchell, Don (2000): *Cultural Geography: A Critical Introduction.* Oxford: Blackwell Publishing.

Reid-Pharr, Robert F. (1994): »Disseminating Heterotopia«, *African American Review* 28 (3), pp. 347–357.

Richards, Dave (2005): »The War at Home: Wood and Burchielli Talk *DMZ*«, *Comic Book Resources* [online], 9 November, http://www.comicbookresources.com/?page=article&old=1&id=6168. 2 December 2011.

Sanchez-del-Valle, Carmina (2011): »*DMZ*'s Dystopic Manhattan: Loving the City, Killing the City«, *International Journal of Comic Art* 13 (1), pp. 529–550.

Soja, Edward W. (1989): *Postmodern Geographies: The Reassertion of Space in Critical Theory.* London: Verso Books.

Wood, Brian (2006): *DMZ Vol. 1: On the Ground.* New York: DC Comics/Vertigo.

— (2007): *DMZ Vol. 2: Body of a Journalist.* New York: DC Comics/Vertigo.

— (2008a): *DMZ Vol. 4: Friendly Fire.* New York: DC Comics/Vertigo.

— (2008b): *DMZ Vol. 5: The Hidden War.* New York: DC Comics/Vertigo.

— (2009a): *DMZ Vol. 6: Blood in the Game.* New York: DC Comics/Vertigo.

— (2009b): *DMZ Vol. 7: War Powers.* New York: DC Comics/Vertigo.

— (2010): *DMZ Vol. 8: Hearts and Minds.* New York: DC Comics/Vertigo.

— (2011): *DMZ Vol. 9: M.I.A.* New York: DC Comics/Vertigo.

— (2012): *DMZ Vol. 11: Free States Rising.* New York: DC Comics/Vertigo.

There's No Place Like Fiction
Narrative Space and Metalepsis in Stephen King's »Umney's Last Case«

JEFF THOSS

Castle Rock, Derry, Jerusalem's Lot—these are well-known names among readers of Stephen King. Over the past forty years, countless of King's horror narratives have been set in these fictional Maine localities. Learning that King's—at the time of writing—latest novel, *11/23/63* (2011), partially takes place in Derry means recalling previous Derry stories, foremost the classic *It* (1986), and imbues readers with an immediate sense of familiarity.[1] Opening the new book, it is as if one were right at home. And should fans wish to stop at Derry or Castle Rock or another location from King's topography of Maine, the author's website hosts a map of the state that allows them to pinpoint the fictional towns among the real ones.[2] Yet to actually visit them is, of course, impossible. It is only in the act of reading, of immersing themselves in the storyworld, that readers can experience King's fictional cosmos.

However, in one of the writer's short stories, the impossible does occur: »Umney's Last Case« (1993) relates how a writer, Samuel Landry, decides to change places with his character, Clyde Umney, substituting his drab present-day reality with the 1930s Los Angeles of a hardboiled detective novel. Throughout the narrative, King conspicuously describes this fictional universe as possessing a clearly delineated and stable space that endows its inhabitants with an equally fixed identity and secure sense of place. The space of reality is, in contrast, repeatedly represented as shifting and essentially unknowable. When Landry, the protagonist, enters the world of his character Umney, these radically different spaces collide and become entangled while identities are confused or lost, even though the ending promises to restore the status quo.

In this contribution, I will employ Jurij Lotman's semiotics of space and the narratological concept of metalepsis to analyze Stephen King's treatment of narrative space in relation to the dichotomy of reality vs. fiction as well as to the issue of escapism that is at the heart of »Umney's Last Case«. As shall be seen, King's narrative confidently highlights the allures of popular fiction's imaginary spaces and readily sets them off against the real world, yet appears somewhat overzealous in reversing and sanctioning the transgression of this border, as if it was trying to cover up the questions it raised.

Before going into the details of »Umney's Last Case«, a few words on Jurij Lotman and metalepsis are in order. According to Lotman, every culture possesses an inherent topological organization that is characterized by a ›binary division‹, a boundary that divides the cultural field and horizontally splits it into two disparate zones. This spatial opposition contrasting an internal with an external field correlates with a semantic opposition; the internal space may be »›cultured‹, ›safe‹, ›harmoniously organized‹,« while the external space may be »›hostile‹, ›dangerous‹, ›chaotic‹« (1990: 131). Lotman lists other topological binaries such as ›up vs. down‹ or ›open vs. closed‹, which may, likewise, be given different political, social, religious, etc. interpretations (1990: 132). In any case, a central place is accorded to the boundary, which »both separates and unites« and »control[s], filter[s] and adapt[s] the external into the internal« (1990: 136; 140)— permanently negotiating the difference between ›cultured‹ and ›hostile‹, for instance. For Lotman, the structure of (narrative) texts mirrors this spatial organization of culture and »becomes a model of [it]« (1977: 217). The typical fairy tale, for example, sharply distinguishes between the safe, closed space of the home and the dangerous, open space of the forest. The boundary between these two zones is essentially impermeable —»only in the forest can terrible and miraculous events take place« (1977: 230)—yet in the course of the story, someone, the hero or heroine, is permitted to transgress it. This »crossing of the basic topological border in the [...] spatial structure« (1977: 238) is what constitutes plot in Lotmanian terms. It is a »revolutionary element in relation to the world picture« (1977: 238) that negates and is superimposed on the purely affirmative and static character of ›plotless‹ texts in which no border-crossing takes place.

In view of Lotman's emphasis on textual boundaries and their transgression, it seems surprising that the connection between his semiotics of space and the concept narrative theory employs to denote the transgression of narrative levels, metalepsis, has so far been underdeveloped. In his specification of Lotman's model, Michael Titzmann argues that the ›modalization of semantic spaces‹ needs to be taken into account; for instance, when a text relates how a character puts into action what they have previously merely fantasized about, the potentially crucial boundary between the spaces of ›fantasy‹ and ›reality‹ may be transgressed (2003: 3082–3083). But while Titzmann here extends Lotman in the direction of modal (or ontological) oppositions and their transgression, he does not yet touch upon metalepsis, which also constitutes an ontological transgression, yet one that is more clearly marked and paradoxical. In its most famous definition, metalepsis is characterized as »any intrusion by the extradiegetic narrator or narratee into the diegetic universe (or by diegetic characters into a metadiegetic universe, etc.), or the inverse« (Genette 1980: 234–235).

Following classical narratology's stratification of narrative structure into story and discourse as well as embedded narratives, a boundary separates these levels that is usually impassable. Characters should not be able to converse with their narrator, nor should they be allowed to disappear into a story someone is telling them. Except, in metaleptic texts, this is what happens, as in Michael Ende's *Die unendliche Geschichte* (1979; translated as *The Neverending Story*), whose hero Bastian enters the book he is reading to save the fictional universe. The spatial implications of such metaleptic transgressions are readily apparent and have been commented upon: different narrative levels, different storyworlds possess different ontologies, different spatiotemporal settings, yet when metalepsis strikes, these are conflated and contaminate one another, resulting in »logically impossible story spaces« (Ryan 2009: 430).

But what of a more distinctly Lotmanian take on the device? Evidently, Lotman is concerned with a boundary that is to be found on the level of the story, a boundary between distinct zones in the same storyworld. He does not directly take into account the boundary between different levels of the narrative (extradiegetic, diegetic, hypodiegetic, etc.) that metalepsis usually involves. Following Titzmann's lead, I hope to show that his analysis can, however, be extended to this type of boundary and that it may behave in a very similar fashion. This, in turn, could benefit both the semiotics of narrative space as well as our understanding of metalepsis.

Even though they need not correspond to the spatial models and metaphors that scholars rely upon to describe narrative structure—one can be further ›up‹ or ›down‹ the hierarchy of narrative levels, while metalepsis may be of an ›ascending‹ or ›descending‹ type—metaleptic texts generally make use of spatial oppositions. Thus, *Die unendliche Geschichte* relies upon an ›outside vs. inside‹ division. Such oppositions may be endowed with non-spatial, symbolic meaning; Ende's novel is well-nigh prototypical here, as ›outside‹ (the ›Outer World‹ in the novel's terminology) corresponds to reality, while ›inside‹ (the world inside the book Bastian is reading) corresponds to the world of unreality, that is, fantasy, fiction, and imagination. In Western culture, the distinction between ›real‹ and ›unreal‹ is, of course, considered to be absolute, and the border between them impassable. Yet metaleptic transgressions—which can themselves only take place in the ›unreal‹ area of fiction—short-circuit the text's separate regions, violate the all-important barrier the narrative has introduced, for instance by having a character move across it. Here one finds Lotman's concept of plot.

To return to Ende's *Unendliche Geschichte*: protagonist Bastian is permitted to move from the ›outside‹ to the ›inside‹, from the actual to the imaginary world, and this is what gives rise to the novel's action and enables it to reflect on the relationship between reality and fiction. In the basic case I have just sketched out, the workings of metalepsis coincide with the spatial structure of narrative as described by Jurij Lotman to a remarkable degree: There is a binary division in the narrative's world based on a spatial opposition overlaid with a semantic one, a theoretically inviolate border between the two zones, and an illicit movement across it that constitutes the crux of the matter. Naturally, there are metalepses that do not fit into this mold, just as not every text that displays Lotman's features may be deemed metaleptic (since it need not involve narrative levels, for example). Nevertheless, in this particular (and popular) narrative pattern, these barriers coalesce and the text shows itself to be just as metaleptic as it is Lotmanian.

One of the popular texts that uses this narrative pattern is »Umney's Last Case«, which not only shares *Die unendliche Geschichte*'s division into a ›real‹ and an ›unreal‹ world, but also deploys the same ›outside vs. inside‹ spatial opposition. The fictional world of private eye Umney is purported to exist »*inside* this man's head« (King 1993/2009: 791; my emphasis), that is, inside Landry's (the writer's) imagination, who must correspondingly be located ›outside‹, in reality. Likewise, Landry's act of transgression, his illicit entry into his character's world, is conceived of as »slip[ping] all the way *in*,« as »plung[ing] all the way *into* the world [he's] created« (1993/2009: 801; 795; my emphasis).

However, in addition to this basic ›outside vs. inside‹ opposition, the narrative also emphatically divides its space into an ›open‹ and a ›closed‹ region, which proves to be an even more significant opposition. The short story begins with a description of Umney's morning walk to the office, which takes the form of a *parcours* through his world, a sequence of familiar landmarks and routine encounters that structure his existence and provide him with a feeling of security. This is made evident in one of the first sentences, in which the recognizable location of a character is seen as signifying universal order: »Peoria Smith, the blind paperboy, was standing in his accustomed place on the corner of Sunset and Laurel, and if that didn't mean God was in His heaven and all was jake with the world, I didn't know what did« (1993/2009: 753). Umney's itinerary continues with a stop at his favorite bar, a chat with the lift operator in his office building, and his concluding arrival at his office, where his secretary awaits him. Movement in the fictional, hypodiegetic universe is circular, a daily visit of the same handful of places, all exuding the comfort and ease of the known. This world is, in fact, literally closed, since it consists of »[j]ust Los Angeles and a few surrounding areas« (1993/2009: 788), as Landry later on tells his character.

The manageable topography of the fictional cosmos is accompanied by a vague yet static temporality, which results in a Bakhtinian chronotope, a fusion of »spatial and temporal indicators [...] into one carefully thought-out, concrete whole« (1981: 84). Umney regularly uses phrases such as »since time out of mind« or »year in and year out« because, as Landry later explains, »time doesn't really pass in this world«, it is always »1930-something« (King 1993/2009: 755; 759; 787; 800). Umney's world is thus not only spatially but also temporally fixed, the clearly delineated space corresponding to a persistent time frame. In other words, the fictional universe forms a closed, hermetically sealed chronotope, as there is neither an outside of L.A. nor an outside of the 1930s that characters could move to. All of this endows characters with an equally static identity and secure sense of place. Peoria will always be the blind paperboy on the corner; there is no provision in the narrative space for a different state of things. And even though Umney travels around in his world, he is restricted to movement »*within* the space assigned to him« (Lotman 1977: 238), which is really a non-movement that continually cancels itself out and returns to the point of departure, thwarting any change in his character and conforming to characteristics of a plotless text.

If the difference between place and space is a conventional association of »security and stability« with the former that is juxtaposed with the »openness, freedom, and threat« of the latter (Tuan 1977: 6), one could say that Umney's fictional universe is all place. Landry's world, the narrative's diegetic universe and real world, is, by contrast, all space. King spends considerably less time detailing this other region of narrative space, though not because it is any less important. This being ›the real world‹, the text could simply expect its readers to already know what it is like, to fill in the textual gaps with their common-sense knowledge. However, I would rather argue that the descriptions of Landry's universe are sparse because it is inscrutable, because it does not offer the clear-cut topography and facile orientation of Umney's world. When the private eye has finally changed places with his author and finds himself on »the other side« (King 1993/2009: 807), he barely dares to leave Landry's home for fear of the unknown that awaits him outside: »Mostly I stay in. I have no urge to explore the world Landry pushed me into« (1993/2009: 810). Hoping to replicate his routine from the fictional world and shut out the real world's volatility and openness, Umney limits himself to a »once-weekly trip to the bank and the grocery store,« yet already »see[s] more than [he] want[s] to« on these outings (1993/2009: 810). In an open world, anything can happen.

In addition, the detective knows that nothing good can befall him in reality from Landry's earlier account of it. According to the writer, this world is characterized by a boundless mutability: after thinking that »life could[n't] get any better« (1993/2009: 795), Landry suffers a reversal of fortune as his son dies of AIDS, his wife commits suicide, and he himself struggles with depression and illness. His narration highlights temporal flux and uncertainty—the proverbial ravages of time—yet this is chronotopically linked with spatial instability. The writer states that, after these traumatic events, »there was only one place left where he could go and feel welcome« (1993/2009: 800)—the fictional universe of the Clyde Umney novels. The time-based upheavals in Landry's life thus correspond to a loss of place in his world, which, in effect, result in a total loss of identity. It is this identity that the writer wishes to regain by switching places with his character, or, rather, it is a different identity, one that is no longer determined by the spatiotemporal chaos of the real world but is constituted by the static spaces and fixed time frame of the fictional one.

This is where metalepsis occurs. Landry moves from the embedding to the embedded narrative level (and Umney subsequently does the reverse), thus violating the boundary between them. It is thus also here where Lotman's plot is to be found, the protagonists crossing the infrangible border between the open and the closed region of narrative space. This transgression is the story's central event—without it, there would be no story—and what happens is more than a straightforward transfer of characters to a different place; the places themselves change. The fictional and the real world spatiotemporally contaminate one another. In order to, as he claims, »prepare« (1993/2009: 790) Umney for the cataclysm that is about to take place, Landry introduces the variability characteristic of his world into the fictional universe he created. The writer »remov[es] all the old landmarks« (1993/2009: 805), causes spatial as well as temporal instability and variation to manifest themselves in Umney's world. Suddenly, the blind paperboy has won the lottery and will undergo a miracle cure for his blindness, the lift operator retires due to cancer, the detective's favorite bar has closed down, his secretary has quit and moved to Oklahoma, and his office building is being renovated.

To the private eye, this transformation of his environment is not only scary but also unfathomable. He deems the transformation »a blasphemy« and reasons that »it meant things would change« though »things weren't supposed to change« (1993/2009: 763; 759). If the frozen topography previously guaranteed his identity, the changes to it now lead him to lose his identity, a fate that Landry had already suffered before he decided to become a character in his own novels. Having lost his footing in the world, Umney can no longer be sure of who he really is. Yet his identity crisis only escalates when he learns that he is but a fiction of Landry's, that his world and the places in it were not real to begin with, and ontological disruption and horror are thus added to the spatiotemporal ones. Just as space constructs and deconstructs the subject, so does metalepsis here, by »breaking down the very structures that apparently define subjects and lend them their air of stability« (Malina 2002: 10). Umney's dissociation from his self is brought to a close when his author reads him his own first-person narration, which ends with the sentence: »So I left town, and as to where I finished up ... well, mister, I think that's my business« (King 1993/2009: 806). Leaving town represents the ultimate impossibility in the fictional universe—there is no outside of »Los Angeles and a few surrounding areas« (1993/2009: 788)—and thus marks its complete collapse.

Naturally, Landry's purpose is not really to destroy the world he created but to take his character's place in it. The text informs readers that he will »tear it apart and rebuild it the way *he* wanted it« (1993/2009: 803), meaning that once Umney has been expulsed, the fictional world will regain its closed and static status, and Landry will have turned into a stable and hence carefree literary character. Or at least, this is the writer's phantasm. As already seen, Umney does not fare well when confronted with the openness of the real world, which largely refuses to be affected by his presence and seems to obey a different rule set. The last part of the narrative, which forms a kind of epilogue, relates how he plots his revenge by becoming an author in his turn and writing himself back into the detective novels. »This time nobody goes home« (1993/2009: 812), Umney states and thereby intimates that the barrier between the text's two zones will be violated one last time—and in the course of this restored—that a final metaleptic transgression is about to occur.

At this point, the story's title, which is also the title of the supposedly final Clyde Umney novel in the diegetic world, reveals its full significance. From Landry's perspective, »Umney's Last Case« aptly fits the tale of his entry into the hardboiled universe and Umney's displacement from it: it is Umney's last case since all future cases will be Landry's. However, the ending intimates that Umney will have one more case: his reentry into his world to dispatch his author. This is potentially a case to end all cases and a case that Landry clearly did not have in mind, leading the title to take on different meanings depending on whether it refers to the story-within-the-story (and the initial metaleptic transgression) or the story as a whole (and the final metaleptic transgression).

The narrative's structure, then, or plot, essentially consists of an opening (double) border crossing (Landry moves from zone A to zone B, Umney from B to A) that is not quite reversed but at least annulled by the potential later border crossing (Umney moves back from A to B). The story is thus somewhat circular, swings back like a pendulum, and while it is not exactly static, it nevertheless drives toward and implies a restoration of the initial state of affairs, an undoing of the, in Lotman's terms, revolutionary element.

Such restitution is in itself not surprising. As Karl Nikolaus Renner explains, the ›principle of consistency‹ governing storytelling demands that contradictions between characters and the space they are located in are resolved (1983: 41–42). There are a variety of ways in which this can be achieved, which can either return the old order (a plot-text that turns out to be actually plotless) or institute a new one (the true plot-text).[3] In »Umney's Last Case«, Landry wishes to cast off his contrary nature in the fictional world by turning into his character, while Umney (vainly) attempts to impose his old spatiotemporal framework upon the new world he finds himself in. In other words, one character attempts to change himself, the other the world around him to do away with existing tensions. If the narrative were to pursue this path, it would possibly lead to the establishment of a new order, but, of course, it instead chooses to append a different, restorative ending. This ending, which reveals the text to be actually plotless, comes about somewhat hastily, almost as an afterthought, as if the narrative was balking at its own themes and was now trying to gloss them over. The precise reasons for this lie in »Umney's Last Case«'s quasi-allegorical qualities and its contradictory stance on escapism.

As has been discussed, the ›inside vs. outside‹ or ›closed vs. open‹ spatial opposition King makes use of is superimposed with the semantic ›fictional vs. real‹ one. Yet one could further argue that the confrontation between the narrative's fictional and real world itself merely sets the stage for a debate on the relationship between popular fiction and reality in general. Umney's crime-novel world, then, becomes the prototype of all of popular culture's storyworlds.[4] Its stasis and rigidity can be seen as exemplifying a traditional type of popular serial storytelling, according to which there is progress within individual episodes—the detective solves a case—but none between them. Each installment finds major characters and settings essentially unchanged and thus permits effortless orientation. Characters do not age in the Clyde Umney detective novels because fiction can simply ignore the passage of time and endlessly revisit the same temporal limbo.[5] Umney's world is narrowly bounded, because storytelling cannot and need not represent the whole world. Instead, it can focus on a small subsection, break things down to a manageable size, and create its own microcosm. The stability and persistency thus achieved stand in contrast not only to highbrow fiction with its supposedly more complex take on these issues, but also to the real world in which time does not stop and spaces cannot be so easily confined.

Needless to say, this is a gross exaggeration and simplification, but it arguably is the way in which popular fiction presents itself in »Umney's Last Case«—as straightforward formula fiction. In any case, it is these qualities that the narrative proclaims to be the »charms« (King 1993/2009: 787) of the fictional world and hence, by extension, of popular storytelling in general. And it is these qualities that it sets off against the menace and hostility of reality, in which everything is transitory and the unknown starts at the threshold of one's front door. With reference to Wolfgang Iser's functionalist distinction between texts that »bolster up the weaknesses of a system [of reality]« and texts that »reveal« them (1978: 83), one could say that popular fiction in »Umney's Last Case« clearly serves the former end; it is designed to conceal and compensate for reality's perceived deficiencies.

If one wants to continue reading the narrative allegorically, as symbolizing the relationship between reality and popular fiction, one is inevitably faced with the question of what the protagonists Umney and Landry represent in the larger context. The detective might pass as the archetypal character to be found in the kind of storyworld just described, yet is Landry meant to be some everyman character, too; the average recipient of popular fiction, for instance? The writer's wish to be his protagonist, his transformation into the hardboiled detective, can be said to illustrate the process of readerly identification with a character, with popular literature offering ready-made identities for those unsatisfied with their own. It also exemplifies escapism, as Landry, who is no longer content with being a »tourist[] in the country of [his] imagination« (King 1993/2009: 795), is not looking for a brief diversion but rather a lasting release from his ordinary existence.

However, even though the text is undoubtedly about escapism, about preferring fictional worlds to the real one, the writer is a rather exceptional escapee. To begin with, Landry is not just any crime fiction reader but the author of the Clyde Umney books, which already makes him special and increases the distance between him and the short story's actual readers. The narrative is also told from Umney's perspective, to whom Landry appears as a nemesis and perennial outsider, thus steering readers' sympathies toward the character and precluding identification with the writer. In addition, Landry's fate is unusually dire: in a short period of time, his son dies and his wife dies and he becomes ill and depressed. By heaping disaster upon disaster in the writer's life, King helps readers empathize and understand why Landry so desperately wishes to escape at the same time as he ensures that they will view his situation as an extraordinary one, one that seems very remote from their own everyday existence.

From »Umney's Last Case«, one gets the impression that it is only those whose sense of self has already been shattered, who have already lost their place in the world, that are prone to long for a different space that will establish and secure their identity. By refusing to make Landry a stand-in for actual readers, the narrative implicitly appears to state that ›normal‹ readers—whoever they may be—could not find the cozy cosmos of popular fiction so appealing as to wish to become a part of it. It also describes a scenario in which it is clearly the individual and not popular fiction that is to be held responsible for the transgression. Although presenting itself as very attractive, the fictional world does not, so to speak, actively lure Landry into its fangs. It is he who initiates the metaleptic place swapping with Umney and who is responsible for (and largely in control of) everything that happens right up until the end.

The narrative thus finds itself in a somewhat paradoxical situation, adumbrating an allegory of escapism yet seemingly circumventing the crux of the matter by representing escapism as the singular fate of an individual, after all. Evidently, it wishes to avoid any self-incriminatory gesture. In the end, the story appears to tell its readers that fiction will get back at you for overstepping its bounds, but unless you are some pathological or traumatized person, you should have no trouble respecting these in the first place. While »Umney's Last Case« deploys metalepsis to negotiate the relationship between fictional and real spaces, it does so to reinforce the boundary between them rather than to obliterate it. Different spatiotemporal settings do get tangled up with one another and enter into a dialogue, yet ultimately the exchange between them is revealed to have been an aberration. The two regions of narrative space only contaminate each other so that readers get a clearer picture of their differences. »Umney's Last Case« posits a stable fictional zone and a dynamic real zone and preserves them as such because, according to its own logic, popular fiction needs this difference in order to function. This is why the story cannot end with Landry and Umney accommodating themselves to their new environment, why Landry must be a dislikable and extreme character designed to prevent identification from the outset, and why a complete restitution is necessary to provide closure. (One can only speculate if all the metaleptic mingling had made King himself somewhat uneasy, had given him the impression that he had gone too far, and thus led him to tag on the more convenient ending.)

If popular fiction is to maintain its appeal, so it appears, it must by needs depict reality as a chaotic place worth fleeing from, as lacking the qualities of order and harmony that are so abundant in fiction. If Umney's world was not so desirable and Landry's world not so flawed, what need would there have been for the writer to even bolster up the weaknesses of his reality with detective novels—a genre which traditionally displays and guarantees the explicability of the world through the solvability of a crime—in the first place? However, there remains a fault line or blind spot in this reasoning: the story's unwillingness to properly address the issue of those who want to disregard or blur the difference between reality and fiction, who wish to permanently cross the border into the land of fantasy. Popular literature, it transpires in King's narrative, requires the opposition between ›closed‹ fiction and ›open‹ reality, and the possibility of a permeable border between the two may very well constitute its true horror.

Notes

1 | I would like to thank Judith Brand and Andreas Mahler for their helpful comments on earlier versions of this piece. My discussion of »Umney's Last Case« is partly based on a chapter from my Ph.D. thesis »Metalepsis in Contemporary Popular Fiction, Film, and Comics« (University of Graz, Austria, 2011), which does not consider space, though.
2 | The map can be found at http://www.stephenking.com/images/map_of_maine.jpg (last accessed 31 July 2012).
3 | For a typology concerning the different possibilities of resolving contradictions, see Renner (1983: 41) and Titzmann (2003: 3080). On the distinction between true plot-texts and actually plotless plot-texts, see Mahler (1998: 8-9).
4 | Of course, many of the attributes thus ascribed to popular fiction apply to literature or storytelling in general, yet a significant number of them also do not, hence the restriction to popular fiction.

5 | A prime example of this process is *The Simpsons* (Fox, 1989–present), whose main cast has not aged or changed in any other significant way in the show's twenty-plus-year run.

References

Bakhtin, M. M. (1981): »Forms of Time and of the Chronotope in the Novel: Notes toward a Historical Poetics«, in *The Dialogic Imagination: Four Essays* (trans. Caryl Emerson/Michael Holquist). Austin: University of Texas Press, pp. 84–258.

Ende, Michael (1979): *Die unendliche Geschichte*. Stuttgart: Thienemann Verlag.

Genette, Gérard (1980): *Narrative Discourse: An Essay in Method* (trans. Jane E. Lewin). Ithaca: Cornell University Press.

Iser, Wolfgang (1978): *The Act of Reading: A Theory of Aesthetic Response*. London: Routledge.

King, Stephen (1993/2009): »Umney's Last Case«, in *Nightmares & Dreamscapes*. New York: Pocket Books, pp. 753–812.

Lotman, Jurij (1977): *The Structure of the Artistic Text* (trans. Gail Lenhoff/Ronald Vroon). Ann Arbor: University of Michigan.

——— (1990): *Universe of the Mind: A Semiotic Theory of Culture* (trans. Ann Shukman). London: Tauris.

Mahler, Andreas (1998): »Welt Modell Theater: Sujetbildung und Sujetwandel im englischen Drama der frühen Neuzeit«, *Poetica* 30, pp. 1–45.

Malina, Debra (2002): *Breaking the Frame: Metalepsis and the Construction of the Subject*. Columbus: Ohio State University Press.

Renner, Karl Nikolaus (1983): *Der Findling—Eine Erzählung von Heinrich von Kleist und ein Film von George Moorse: Prinzipien einer adäquaten Wiedergabe narrativer Strukturen*. Munich: Wilhelm Fink Verlag.

Ryan, Marie-Laure (2009): »Space«, in Peter Hühn et al. (eds.), *Handbook of Narratology*. Berlin: Walter de Gruyter, pp. 420–433.

Titzmann, Michael (2003): »Semiotische Aspekte der Literaturwissenschaft«, Roland Posner/Klaus Robering/Thomas A. Sebeok (eds.), *Semiotik/Semiotics: Ein Handbuch zu den zeichentheoretischen Grundlagen von Natur und Kultur*, Vol. 13.3. Berlin: De Gruyter, pp. 3028–3103.

Tuan, Yi-Fu (1977): *Space and Place: The Perspective of Experience*. Minneapolis: University of Minnesota Press.

The Black Hole at the Heart of America?
Space, Family, and the Black Hallway in *House of Leaves*

MICHAEL FUCHS

When *House of Leaves* hit bookstores in March 2000 after years in the making (and partial online pre-publication since 1996), it not only quickly became a bestseller, but also turned into one of critics' favorite pieces of fiction that year. *The Guardian*'s Steven Poole, for example, wrote that the book was »a superbly inventive creation« that was »genuinely exciting in its technical and literary exuberance« (2000: par. 8; par. 11), while *New York Times* critic Robert Kelly similarly praised *House of Leaves* as being »funny, moving, sexy, beautifully told, [and] an elaborate engagement with the shape and meaning of narrative« (2000: par. 2).

Even though one would think that literature scholars from around the world would be eager to investigate (and publish on) an experimental masterpiece like *House of Leaves*, its author, Mark Z. Danielewski, has complicated the process of »academic onanism«[1] (Danielewski 2000: 467). After all, *House of Leaves* consists of blind Zampanò's ekphrastic descriptions and his (more or less) academic treatment of the non-existing video *The Navidson Record*,[2] filmed by Pulitzer Prize-winning photojournalist Will ›Navy‹ Navidson. *The Navidson Record* depicts the strange events occurring in the Navidson family's[3] (consisting of Will Navidson, Karen Green, and their children Chad and Daisy) new house in Virginia and appears to be a documentary, but Zampanò believes it to have been staged. Zampanò's manuscript is edited and commented on by Johnny Truant, who happens to discover Zampanò's fragmentary notes after his death and becomes so obsessed with the text (and the video[4]) that he increasingly inhabits a borderland between reality and fiction. These personal narratives surrounding Johnny's increasing obsession with completing the book are collected in his footnotes to Zampanò's manuscript pages. Yet Johnny's notes are not the final narrative layer found in *House of Leaves*, as they are, in turn, framed by remarks and notes by anonymous editors, who are also presented as the authors of the short foreword to *House of Leaves*. In addition, the book[5] features three appendi-

Illustration 1: House of Leaves' *narrative levels*[6]

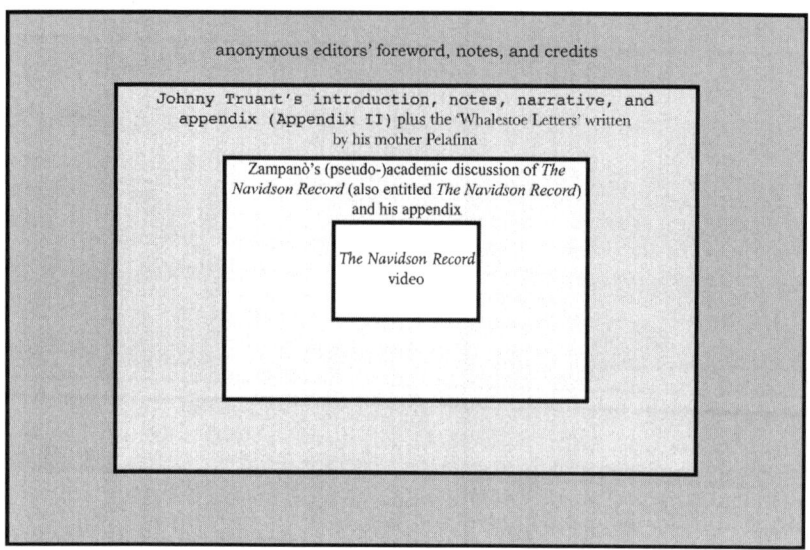

Source: Author's illustration.

ces collected by the three main narrators (Zampanò, Truant, and the editors) and an ›index‹ assembled by the anonymous editors, which provides partly useful entries, but also nonsensical ones (for instance, both ›and‹ and ›in‹ are listed), not to mention the fact that the cross-references to specific pages are at times entirely wrong (see *Illustration 1* for a diagram of *House of Leaves*' narrative layers).

This basic construction of narrating *House of Leaves* not in the form of a traditional haunted house tale, but as a (more or less) scholarly commentary on a video displaying an actual haunted house accompanied by the musings of several narrators that are at times closer and at other times farther away from the haunted house tale that is at the book's core, allows Danielewski to foreclose academic commentary on *House of Leaves* to a certain extent. Nele Bemong is among those who have fallen into the traps of this meta-commentary on the text within itself. She opens one of the earliest scholarly essays on *House of Leaves* by rightfully stressing that »Danielewski *seems* to make the task of the literary theorist redundant« (my emphasis), but then goes on to argue that scholars should »not lose track of the fact that this theoretical discourse is just as much an essential part of the book, and thus object of our study, as the story itself« (2003: par. 1). Yet Bemong's psychoanalytical reading that centers on the uncanny ends up quoting large chunks of secondary sources that are already cited in *House of Leaves* to analyze the events depicted in *The Navidson Record*, leaving one wondering what her commentary actually adds to the understanding of her object of study, which supposedly also includes Zampanò's secondary texts.[7]

More importantly for the present purposes, the book's structure already indicates the importance of different layers of ›realities‹, different spatio-temporal dimensions, different spaces to making sense of *House of Leaves*. Questions related to space are omnipresent not only in the book, but also in *House of Leaves* scholarship.[8] Alison Gibbons, for example, has discussed the »iconic spatial arrangement« (2010: 294) of words on certain pages in the book, as verbal signs visually recreate narrative action. In addition, Gibbons devotes part of her essay to the intermedial spaces occupied by a multimodal book such as *House of Leaves*, an issue that has also been taken up by Christian Zolles (2011). Steven Belletto (2009) has discussed a similar interstitial location, namely the book's placement on the borders between fictional narrative and factual academic treatment, and Rune Graulund (2006) has published on the problematic nature of trying to locate the borderlines between *House of Leaves*' text and paratexts. William G. Little struggles with classifying the book »as a postmodern novel or as something clearly other than a postmodern novel (a post-postmodern novel?)« only to conclude that »such a placement project« cannot but lead »to a dead end« (2007: 170), while Katharine Cox (2006) has taken an entirely different approach and investigated the places and spaces occupied by the feminine in the book. In addition, *House of Leaves*' labyrinthine structure has garnered quite some academic interest (Dupuy 2008; Hilton 2008; Gehring 2009; Message 2009), as has its transgression of the usual spatial limits of a book, since, in many ways, it rather resembles a piece of hyperfiction than print fiction (Chanen 2007; Hansen 2004; Pressman 2006). To these topics, one might add the repeated metaleptic transgressions of (seemingly) ontologically distinct realities and the fact that *House of Leaves* doesn't even observe the material limits of the book page (see *Illustration 2*).[9]

It is not highly surprising that questions related to spatiality feature so prominently in *House of Leaves* scholarship given that even Johnny Truant realizes that, at the end of the day, *The Navidson Record* primarily is »about space«, while Zampanò highlights that »[i]t is impossible to [overestimate] the importance of space« (Danielewski 2000: xix; 41) in the documentary that lies at the heart of the book. And these two statements by *House of Leaves*' primary narrators are merely two of numerous examples that self-reflexively equate the film in the book with the book, in this case underscoring that like *The Navidson Record*, *House of Leaves* is all about space. In so doing, *House of Leaves* shows that, in many ways, the book is about America. After all, like *House of Leaves*, America is all about space, for American history and identity have been intricately tied to issues of spatiality, from the ›City Upon a Hill‹ to the Space Race, and from the dislocation of Native Americans to the increasing atomization of American society, not to mention the transnational (soft) power of the United States. Indeed, space is so closely linked to conceptualizations of ›Americanness‹ that Charles Olson exclaimed »SPACE to be the central fact to man born in America« (1947/1997: 11).

Illustration 2: House of Leaves *repeatedly breaks the spatial confines of the book page.*

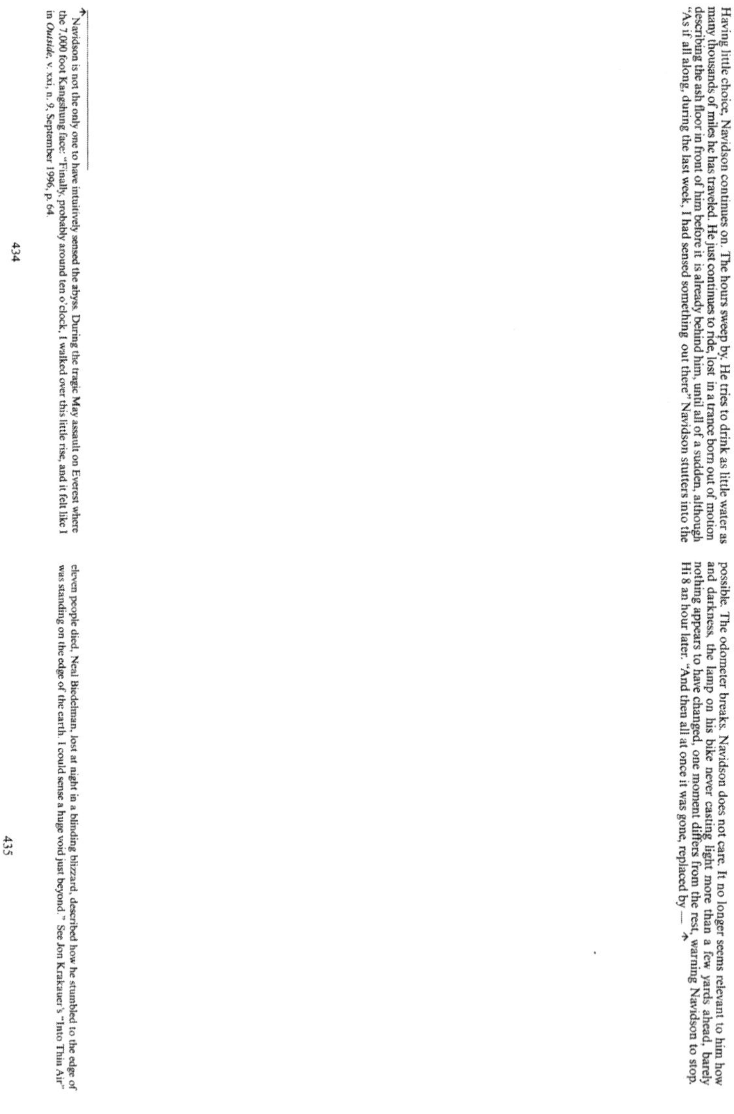

Having little choice, Navidson continues on. The hours sweep by. He tries to drink as little water as possible. The odometer breaks. Navidson does not care. It no longer seems relevant to him how many thousands of miles he has traveled. He just continues to ride, lost in a trance born out of motion and darkness, the lamp on his bike never casting light more than a *few* yards ahead, barely describing the ash floor in front of him before it is already behind him, until all of a sudden, although nothing appears to have changed, one moment differs from the rest, warning Navidson to stop. "As if all along, during the last week, I had sensed something out there" Navidson stutters into the Hi 8 an hour later. "And then all at once it was gone, replaced by — ↑

↑Navidson is not the only one to have intuitively sensed the abyss. During the tragic May assault on Everest where eleven people died, Neal Biedelman, lost at night in a blinding blizzard, described how he stumbled to the edge of the 7,000 foot Kangshung face: "Finally, probably around ten o'clock, I walked over this little rise, and it felt like I was standing on the edge of the earth. I could sense a huge void just beyond." See Jon Krakauer's "Into Thin Air" in *Outside*, v. xxi, n. 9, September 1996, p. 64.

434 435

Source: Mark Z. Danielewski, *House of Leaves*, pp. 434-435. *House of Leaves* © Mark Z. Danielewski, 2000.

As Klaus Benesch thus concludes in his introduction to *Space in America*, »[i]f American democracy, as Frederick Jackson Turner claimed, ›is born of free land‹, then its history may well be defined as the history of the fierce struggles to gain and maintain power over both the geographical, social, and political spaces of America and its concomitant narratives« (2005: 19). Since *House of Leaves* zeroes in on the question of how representations shape our world(view)s, it is hardly surprising that its discussion of America centers on the narratives that, in fact, make ›America‹, for ›America‹ is more of an idea than an actual geographic place—»the idea that it is the realization of everything others have dreamt of— justice, abundance, rule of law, wealth, freedom« (Baudrillard 1986: 77).[10]

Before, however, delving into the intricacies of space, especially distinctly American spaces, in *House of Leaves*, a warning may be due: When discussing a book that strongly capitalizes on the interactive (or, should it rather be, ›interpassive‹[11]?) potential of the medium, a ›writerly text‹ (Barthes 1970) if there ever was one, the preformed and performed reading protocols each recipient carries into his/her encounter with *House of Leaves* decidedly affect not merely the individual reading experience, but, even more so, the actualization of meaning potentials. As a result, if you, dear reader, expect a close reading that cracks the various codes spread throughout the book, clarifies its numerous inconsistencies, and/or explains its incoherencies, then, to borrow from *House of Leaves*, this chapter is not for you. Furthermore, the fact that my following musings will focus on the ›American‹ narrative surrounding the Navidson family may cast the shadows of incoherency and inconsistency aside to a degree that already misrepresents *House of Leaves* in many ways.

AMERICAN HAUNTINGS

In his writings on *The Navidson Record*, Zampanò, states that the video is frequently »catalogued as a gothic tale« (Danielewski 2000: 3). Since the events unfolding in the documentary (and its effects on Truant) take center stage in *House of Leaves*, the book can be assigned to the same genre, that of gothic fiction. However, *House of Leaves* is not merely a piece of gothic fiction, but decidedly a successor to the tradition of the American Gothic.[12] As Teresa A. Goddu notes, »the American Gothic consists of a less coherent set of conventions« (1997: 4) than its European counterpart.[13] This heterogeneity can, of course, be linked to the diversity of the American nation and the democratic fissures resulting from the white myths that (still) dominate the American imagination. From such a postcolonial vantage point, Jeanette Idiart and Jennifer Schulz argue that

American Gothic literature reflects the ›haunted consciousness‹ of the nation: the awareness that at the heart of its governing text are contradictions that threaten to unveil American democracy as a fiction. These texts bring to the surface the knowledge that the founding value of the Constitution, equality, is predicated on the exclusion of selected populations on the basis of race, gender and property. (1999: 127)

Helen Wheatley thus concludes that »the political identity of the United States, questions of national guilt and conspiracy, the treatment of Native American communities, the legacy of slavery and, more latterly, American foreign policy, dominate our understanding of the American Gothic« (2006: 123). *House of Leaves* critically engages with some of these questions, most importantly American foreign policy and the removals and slaughter of Native Americans.

In terms of American foreign policy, Will Navidson arguably represents America. A comparison with his twin brother Tom highlights Navy's inherent ›Americanness‹: »Tom just wants to be, Navidson must become« (Danielewski 2000: 32), thus underlining Will as the embodiment of the future orientation so defining of the United States. He is furthermore described as a »classic hunter« (2000: 37), thereby intricately connecting Navidson to one of America's most enduring national myths, that of the frontier (more on that in the next section of this chapter). In his role as a ›true‹ American, Will Navidson has toured the world as a photojournalist. Little do readers get to know of Navy's past journeys that seemingly took him to the remotest locales on the globe; however, bits and pieces of information about his Pulitzer Prize-winning photograph surface. As these details are uncovered, it becomes clear that Will is haunted by guilt resulting from shooting the photo rather than intervening in the realities of Third World poverty.

Navy's prize-winning photograph was shot in Sudan, where Will had to »walk the violent« and »disease-infected streets« to »finally discover [a] child on some rocky patch of earth« (2000: 419).

In the photograph, [a] vulture sits behind Delial, frame left, slightly out of focus, primary feathers beginning to feel the air as it prepares for flight. Near the centre, in crisp focus, squats Delial, bone dangling in her tawny almost inhuman fingers, her lips a crawl of insects, her eyes swollen with sand. Illness and hunger are on her but Death is still a few paces behind, perched on a rocky mound, talons fully extended, black eyes focused on Famine's daughter. (Danielewski 2000: 420)[14]

This image of the girl whom Navidson named ›Delial‹ should start to haunt him later in his life, as he began to wonder why he wasn't »just doing something about this instead of just photographing it« (2000: 394). While the photograph's near omnipresence strongly reflects the girl's absence, in a rather imperialist move, his appellation, ›Delial‹, confers neither voice nor identity to the girl, marginalizing her in the process in favor of the photograph per se.

The specter of Delial takes full force in a letter drunken Navidson writes to his companion Karen:

> I can't get Delial out of my head. Delial, Delial, Delial—the name I gave to the girl in the photo that won me all the fame and gory, that's all she is Karen, just the photo. [...] But the photo, that's not what I can't get out of my head right now. Not the photo—that photo, that thing—but who she was before one-sixtieth of a second sliced her out of thin air and won me the pulitzer though that didnt keep the vultures away i did that by swinging my tripod around though that didnt keep her from dyding five years old daisy's age except she was pciking at a bone you should have seen her not the but her a little girl squatting in a field of rock dangling a bone between her finger i miss miss miss but i didn't miss i got her along with the vulture in the background when the real vulture was the guy with the camera preying on her for his fuck pulitzer prize. (Danielewski 2000: 391-392)

Obviously, Navidson is less haunted by the memory of Delial per se than his feeling of guilt, as his ›haunted consciousness‹ about his exploitative deeds catches up with him. If, as I've suggested above, Will represents the United States, then Navy's actions mirror the international activities by the U.S. (Navidson's nickname, of course, reinforces such a perception). And, indeed, one cannot help but find similarities, not only in terms of how Navy is but one media agent exploiting Third World tragedy for fame, increased television ratings, or a higher print run, but also in terms of how U.S. foreign policy for a long time has only intervened in international conflicts when it was to its own advantage and supported its dominant role as world power number one. That these goals have often been achieved by subduing those who are already worse off seems to be part of the game.

From a poststructuralist perspective, this tale about the exploitation of underprivileged persons and countries from the Third World, about the global shadow cast by the pursuit of the American Dream, can easily be transformed into an allegory about internal U.S.-American struggles, most importantly the disenfranchisement of indigenous peoples. In *House of Leaves*, such a metonymic extension is, however, thwarted by an explicit break with a distinctly American haunted house tradition. After Karen has told realtor Alicia Rosenbaum about the events that unfold in the house, Alicia digs into the history of the house, admitting that she hopes »to find some kind of ghostliness« (2000: 409). When she can't discover anything worryingly out of the ordinary, she goes »ahead and check[s] if the house was built on an old Indian burial ground« (2000: 409), thus tapping into a tradition especially known from movies such as *The Amityville Horror* (1979), in which the spirits of slaughtered Native Americans rise and enact their revenge on white families, exorcizing white guilt for practically turning the entire American continent into an Indian burial ground in the process. However, as Alicia assures Karen, this is »definitely not« the case with the Navidsons' new home, for »[i]t's all too marshy with winter rains and the James River nearby,« which makes the area »[n]ot a good place for a cemetery« (2000: 409). Yet Alicia highlights that »the only thing distinguished about [the Navidsons'] home's past« (2000: 409) is the communal past of Virginia, most importantly the beginnings of English settlements surrounding Jamestown. Tellingly, Zampanò traces the existence of the haunted house at least as far back as 1610, when three early British settlers went searching for game but only »found ftaires« (2000: 414), that is, stairs that seemingly led everywhere and nowhere at the same time, like those discovered in the Navidsons' new home about 380 years later.

This explicit emphasis on Virginia and Jamestown offsets the earlier negation of the specter of Native American annihilation looming large over the house in Ash Tree Lane, *The Navidson Record*, and *House of Leaves* to a certain extent. Jamestown, after all, is the place where one of the great American myths was born, that of Native American queen Pocahontas saving Captain John Smith. After marrying John Rolfe, Pocahontas was baptized out of her own will a couple of years later and chose the name Rebecca, a name that should become uncannily prophetic considering a bible verse that reads, »And the Lord said unto [Rebecca], ›Two nations are in thy womb, and two manner of people shall be separated from thy body; and the one people shall be stronger than the other people‹« (KJ21, Gen 25: 23).[15] Rebecca's people had practically vanished 75 years later, exterminated by the ›stronger‹ white colonists.

In addition to the significance of Jamestown, the naming of ›Virginia‹ alone, the virgin land to be explored and conquered by Anglo-Saxon settlers, also calls to mind the troubled past of the United States.[16] This idea is, of course, inherently connected with the notion of Manifest Destiny, the belief that the white race was divinely destined to settle across the continent, which cannot be discussed without considering the context of the frontier myth.

RE-DEFINING THE FRONTIER: EXPLORING HOME

In his now famous presentation at the 1893 meeting of the American Historical Association, Frederick Jackson Turner maintained that the frontier experience was *the* central force in creating an American identity. »The existence of an area of free land, its continuous recession, and the advance of American settlement westward, explain American development« (1920/1986: 1), he argued. And countless pieces of popular culture, from the *Star Trek* universe to video games such as *Red Dead Redemption* (2010), have kept the central ideas and ideals connected to the frontier myth in circulation to this day, assuring its »powerful continuing presence in contemporary culture« (Slotkin 1985: 15).

The frontier has always been conceived as an imaginative border between the safety and security of ›civilization‹ and the dangers of untamed nature,[17] the place where white and other (primarily Native American) cultures clash, and the site where the known meets the unknown. Especially the latter point provides a welcome bridge to Gothicism. After all, Sigmund Freud described the uncanny, one of the characteristic features of gothic literature, as a curious blend of the familiar and the unfamiliar: »the uncanny is that class of the frightening which leads back to what is known of old and long familiar« (1919/1955: 220). In *House of Leaves*, gothic and frontier traditions merge, as not only the new Navidson home suddenly turns from secure and ›homely‹ to threatening and ›unhomely‹, but a dark hallway emerges in the house that relocates the frontier experience into the middle of an American family home.[18]

The Navidson Record opens with a clear declaration of Navy's intentions:

I just want to create a record of how Karen and I bought a small house in the country and moved into it with our children. Sort of see how everything turns out. No gunfire, famine, or flies. Just lots of toothpaste, gardening and people stuff. [...] I just thought it would be nice to see how people move into a place and start to inhabit it. Settle in, maybe put down roots, interact, hopefully understand each other a little better. Personally, I just want to create a cozy little outpost for me and my family. (Danielewski 2000: 8-9)

In this introduction, Navy clearly differentiates between the dangers of the outside—›gunfire, famine, or flies‹—and the safety of the home. A couple of pages later, Zampanò even highlights the use of ›outpost‹ in the film's opening, writing that »[b]y definition ›outpost‹ means a base, military or other, which while safe inside functions principally to provide protection from hostile forces found on the outside« (2000: 23). However, as the family has to realize rather soon, they need as much shelter against the inside as against the outside.

The mystery surrounding the house begins to unfold when the family returns from a wedding in Seattle, only a couple of weeks after having moved into the house in Ash Tree Lane. »When they returned, something in the house had changed« (2000: 24). While, at first, it remains unclear what exactly happened (indeed, one can never really say what *exactly* happened), »there had been an intrusion« into the (seemingly) secure home, and »the change [...] destroyed any sense of security or well-being« (2000: 24; 28). A fundamental anxiety emerges in the family—they no longer feel »quite at home in [their] own home« (Vidler 1992: 3-4).

As they look for signs of intrusion, the family is utterly surprised to find a new room separating the parents' and the children's rooms; a strange occurrence, indeed, but one that can be explained rather easily, for the house's blueprints »confirm the existence of a strange crawl space [...] running between both bedrooms« (Danielewski 2000: 29). However, the sudden appearance of the new room is only the beginning, for Navy soon discovers that »the width of the house inside [...] exceed[s] the width of the house as measured from the outside by 1/4"« (2000: 30), as if, in an allegorical inversion, he realized that all the utopian ideas projected onto America from the outside couldn't be fulfilled on the inside. Flabbergasted, Navidson asks his twin brother Tom (who he had hardly been talking to for eight years) for help. Tom, however, cannot do much other than establish that »[t]he interior of the house exceeds the exterior not by 1/4" but by 5/16"« (2000: 32). Since both brothers still cannot explain what to them seems to be an error, Navidson calls an old friend, Billy Reston, who teaches engineering at the University of Virginia. Billy arrives with state of the art equipment, and the scientific authority emanating from the tools creates a sense of security. Only momentarily, though, because just when »[i]t appears the discrepancy has finally been eliminated« (2000: 39), Karen notices that one of the house's walls has moved by more than a foot. These are the first signs that »the language of objectivity can never adequately address the reality of that place on Ash Street Lane« (2000: 378–379). However, the malleable shape of the house provides only a first taste of the house's surprises. The real horror starts to unfold when a black hallway suddenly appears in the living room several weeks later.

The mere presence of the hallway awakens a dormant side within Navy, for he »has always been an adventurer willing to risk his personal safety in the name of achievement« (2000: 60). Karen succeeds in keeping Will's urges to venture into the black unknown in check for some time. Yet when a pair of unnamed celebrities seemingly unexpectedly drops in for dinner one day and disappears inside the house, Navy finally gives in to his desire to explore that unknown presence in his home and to penetrate the house's ›heart of darkness‹. As he, unbeknownst to Karen, moves »deeper and deeper into the house,« Navidson only notices »the otherness inherent in that place« (2000: 64). The Navidsons decide to ask for help and hire the »professional hunter and explorer« (2000: 80) Holloway Roberts (who recruits Jed Leeder and Kirby ›Wax‹ Hook to help in his endeavors) to explore their home. When he arrives at the house, he »looks less like a guest and more like some conquistador landing on new shores« (2000: 80), making his characterization as the classic frontiersman explicit.

It is noteworthy that while the men struggle with that which cannot be explained and employ various tools in order to solve the (unsolvable) mysteries posed by the house, »the children [...] just accepted [them]« (2000: 39). Karen, on the other hand, »challenges [the house's] irregularity by introducing normalcy« and »remains the standard bearer of responsibility and is categorically against risks especially those which might endanger her family or her happiness« (2000: 37; 60). As Richard W. Slatta writes, »[t]raditional frontier myths' focus has a decidedly masculine flavour—the mountain man, the intrepid explorer, the lone cowboy or gunman—rugged individualists all« (2010: 88). With the exception of Tom, *House of Leaves'* male characters are cut from the same cloth. When Roberts enters the house, Navidson finally realizes that »he must ask another man to explore his own house,« and a »classic male struggle for dominance« ensues (Danielewski 2000: 82; 85). This struggle, however, not merely pits Navidson against Roberts, but, even more so, both of them against the unconquered nothingness that awaits them in the hallway. Essentially, their explorations into the dark void looming in the middle of the Navidson home are nothing but »attempt[s] to territorialize and thus preside over that virtually unfathomable space« (2000: 386), only to realize that there is nothing there to territorialize. As is the case with more or less any party venturing into the unknown, trying to expand the known world, the team exploring the Navidson home has to deal with losses—Holloway disappears (and probably commits suicide), Wax is severely injured, Jed is killed, Tom vanishes in the darkness, and Navidson is also physically scarred for life, as they all have to confront their personal demons during their respective journeys into the unknown.

By re-configuring the frontier as being located in the home, *House of Leaves* recontextualizes the frontier experience as an encounter with the self more so than the other. In addition, *House of Leaves* provides testament to Henry James' old adage that »a good ghost story [...] must be connected at a hundred points with the common objects of life« (1865/1984: 742). More recently, academics have similarly emphasized the importance of domestic life to the American Gothic. Leading Gothic scholar Fred Botting, for example, writes that »the bourgeois family is the scene of ghostly return, where guilty secrets of past transgression and uncertain class origins are the sources of anxiety« and adds that »the commonplace of American culture was full of little mysteries and guilty secrets from communal and family pasts« (1996: 114–115).

INVESTIGATING THE AMERICAN FAMILY

Indeed, from Charles Brockden Brown's *Wieland* (1798) and Edgar Allan Poe's »The Fall of the House of Usher« (1834) to more contemporary examples such as Stephen King's *Pet Sematary* (1983) and movies such as *Poltergeist* (1982) and *Insidious* (2011), gothic horror has always centered on family melodrama. Richard Davenport-Hines rightfully argues that this omnipresence of family issues in the American Gothic results from the nation's idealization of the nuclear family: »[A]s Americans adopted a specialised, even extremist veneration of family, some of their writers adapted Gothic imagery to exemplify the destructive power of families. Gothic excess was deployed to represent domesticity's extreme horrors« (1998: 267).[19] Scholars have especially highlighted the destructive powers of middle- and upper-class families, who »[a]s [...] institutional prop[s] of bourgeois capitalism [...] produc[e] colonized subjects and reproduc[e] ideological values,« because »vulnerable individuals« are turned into »neurotic conformists« (Williams 1996: 14–15). These approaches usually rely on Freudian psychoanalysis. Freud argued that civilization relies on repression, but that which it represses tends to resurface, usually in distorted form as a violent reaction against agents of repression, as the self seeks access to the repressed, from which it is »being blocked off« but »to which it ought normally to have access« (Sedgwick 1980: 13). And it is repression that *The Navidson Record*, like so many Gothic tales before, essentially revolves around.

Already early on, readers get to know that the Navidson family is not as perfect as it might seem from the outside. On the outside, there's Pulitzer Prize-winning photojournalist Will Navidson, his long-time companion and former model Karen Green, their statistically appropriate two children, boy Chad, eight, and girl Daisy, five, plus a male Siberian husky named Hillary and tabby cat Mallory. On the inside, the relationship between Navy and Karen is »foundering« and characterized by »increased alienation« that is also due to »eleven years of constant departures and brief returns« on Navy's part (Danielewski 2000: 10). Confronted with the choice between career and family life, Will »compromises by turning reconciliation into a subject for documentation« (2000: 10), writes Zampanò. However, performatively speaking, Navy's move into the new home in Virginia marks his entrance into domestic family life and the conclusion of his career as adventurous photojournalist. The words in his introduction to *The Navidson Record* quoted in the previous section show that Will clearly regards family life and his career as opposing forces, and that he willingly chooses the security of domestic life over the risks of his life as a photojournalist. However, the ensuing events overturn any intentions of new familial unity and of »bring[ing] them closer together« (2000: 82) rapidly.

Tellingly, the first uncanny event occurring in the house, as mentioned above, is the appearance of a new room between the parents' and the children's rooms. Where once was a wall separating them, there suddenly are two doors that lead into a small closet-like space connecting the two rooms. While, at first, the doors could be regarded as a bridge connecting parents and children, one should not forget that the doors don't directly lead into the adjacent room, but that a new room is now located in-between, increasing the physical space between parents and children, which, of course, signifies that the house begins to tear the family apart. Within days, the children are ordered out of the house and »drift farther and farther out into the neighborhood for increasingly long spates of time« (2000: 56).

Though one might expect the family to get closer in face of the emerging danger, the »invasions begin to strip the Navidsons of any existing cohesion« (2000: 83). Most importantly, the physical changes in the house awaken the adventurer in Will and the homely housewife in Karen. For Navy, the house simply can no longer represent safety and domestic life. The ensuing explorations of and into the house generate even more tensions.

During the explorations, especially the children are in distress. Even though, as mentioned earlier, they seemingly ›just accepted‹ the strange events around them in the beginning, Chad and Daisy are increasingly worried. Their drawings provide glimpses into the children's repressed fears, as they are filled with »shallow lines and imperfect shapes narrating the light seeping away from their lives« (2000: 315). Their parents, however, hardly react to the obvious signals. The night Navidson, Tom, and Reston start a rescue mission to retrieve Holloway, Jed, and Wax from the dark unknown, »Chad and Daisy have to put themselves to sleep« (2000: 315). At long last, Karen »snaps out of her obsession« and thoughts about Navy's well-being when the »children come racing down into the living room, claiming to have heard voices« (2000: 315).

Tellingly, when Tom and Reston bring bloodied Wax and dead Jed (both have been shot by Holloway) back into the house (ok, technically, they never left the house) about forty hours later, the children are left alone with Wax and Jed in the kitchen. One of Chad's teachers, Teppet Brookes, suspects domestic violence due to Chad's dark paintings and decides to pay the family a visit. The potential heroine who wants to rescue the children from an unsafe environment enters the house just in time to see Jed and Wax emerging from the hallway. She »fails to utter even one word or offer any sort of assistance« and leaves rather quickly, and »the children are once again abandoned« (2000: 317; 319).

A note (*Illustration 3*) in Truant's appendix highlights the aim, or, maybe, rather wish, to condemn Navy and Karen for ignoring their children in face of epistemological and ontological uncertainty and rather give in to detachment not only from their children but their entire family. In the book, the majority of the hand-written parts are in red ink and the note is part of a collage of three text fragments that touch upon many of the guiding themes—the labyrinthine structure of the book, the haunted house, and parental narcissism—and the merger of the verbal and the visual in the book. More importantly, however, the note is presented as a possible path the narrative could have taken that is excluded from the main narrative but still included in the book. The note depicts the desired but unfulfilled plot—something repressed, but obviously surfacing in the appendix—that features the children as sacrificial lambs. While the imagined narrative arch would most likely end with the sacrifice and the loss of the two children, both children survive in the book. Chad escapes from the house on his own after hearing »the sound of a dying man« (2000: 320), and Daisy is saved by Tom, who sacrifices his own life in the process, paving the path for Will to step into the dark hallway again several weeks later to search for his lost brother. Although one cannot say how the ›alternative‹ plot might have unfolded, it seems as if it would have taken a much more sinister twist than the one narrated in *House of Leaves*.[20]

Illustration 3: One of Zampanò's notes outlines a path the narrative of The Navidson Record *could have taken*

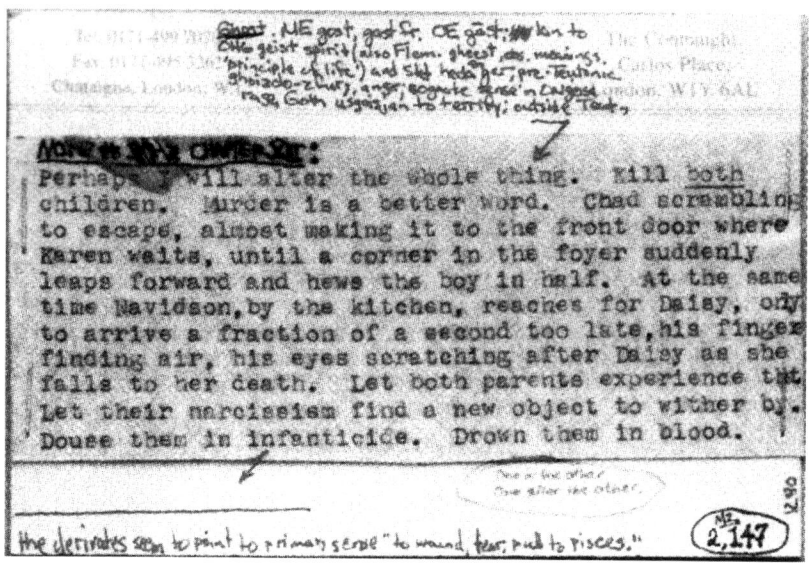

Source: Mark Z. Danielewski, *House of Leaves*, p. 552. *House of Leaves* © Mark Z. Danielewski, 2000.

In the narrative path depicted in the main plot, Karen is not punished, but rather eventually overcomes her repressed memories of being sexually abused as a child and thus her fear of intimacy.[21] Indeed, she literally surmounts the distance between herself and Navy, steps into the dark hallway and thus penetrates the visible, yet symbolically invisible, border separating them, which allows both of them to reunite with their children, too. The final, and rather short, chapter of the book (and the documentary) depicts the Navidson family (with Will and Karen married) in their new home in Vermont. Will's final voyage into the dark unknown has left him physically scarred,[22] Karen has to undergo mastectomy and chemotherapy some time after the events in the house on Ash Tree Lane, and the Navidsons supposedly »will never be able to leave the memory of that place« (2000: 526). The »warmly lit shots« with which Navidson captures their new life and the »[h]undreds of photographs hang[ing] on the walls of their home« (2000: 527) draw an incredibly positive picture of family life, as if Navidson and Karen have overcome the psychological and physical distance between them and finally settled in. Yet on another level, the extremely positive images surrounding them in their new home become active agents in the attempt to, indeed, forget about the events that took place in Virginia, again granting power to the destructive force that is repression.

THE MEANINGS OF NOTHINGNESS

Zampanò and numerous of his sources repeatedly stress that »the hallway remains meaningless« (2000: 60). But does it remain meaningless? Without much thought, one could easily argue that the dark hallway, that void that tellingly appears in the midst of the living room of a more than typical American family—father, mother, son, daughter, cat, and dog (albeit mom and dad are not married)—signifies how the dark and guilt-ridden past (or maybe even the sanctification and/or forgetting of the past) of the United States is at the core of the nation. Indeed, in general, one can hardly escape the feeling that the core ideals and myths that America was founded on are largely being dismantled in the book.[23]

Yet despite the criticism and subversive stance toward the great American myths, the metanarratives that unite the ›imagined community‹ (Anderson 1983/2006) of Americans, the book's end (temporally speaking; its beginning from a structural perspective) finds Johnny in Hollywood, California, after having journeyed eastward to trace a trip Zampanò had undertaken. Truant thus follows his manifest destiny and ›goes west‹ until reaching Hollywood, the locale where the film industry manufactures not merely dreams, but the American Dream,[24] to complete *House of Leaves* and thus reach the goal he has become so obsessed with. Indeed, the presence of the American Dream cannot be denied throughout the novel, as Navy is haunted by memories of Delial, the starving girl who Will exploited for the photograph that brought him the Pulitzer Prize—a dire image of imperialism combined with cut-throat capitalism that leaves a bittersweet aftertaste to the Darwinian notion of the ›survival of the fittest‹ that is not only so frequently used in business circles but which Navy has also participated in on his journey »to gain a great deal of fame and fortune« (2000: 91). While exploring the house, Holloway, on the other hand, is on an »epic progress« toward »joining Navidson's world, [which] he perceives as a place for the esteemed, secure, and remembered« (2000: 91; 98). And if there is a geographical location assigned to that very place in the American imagination, it is Hollywood, which, much like America itself, is more of an idea than an actual geographical location in the state of California. However, whether the fact that Truant writes his final words at that storied place indicates that he is about to join (or has already joined) the ranks of the esteemed and remembered, remains pure speculation.

More tellingly, the introduction is dated October 31,[25] thus turning *House of Leaves* into Johnny's contribution not to mere Halloween traditions, but also to the traditions of the American Gothic. When Karen first sees the dark hallway in the living room, she »freezes on the threshold« (2000: 57). The threshold turns into a space she, effectively, »cannot penetrate« (2000: 84) until she wants to reunite her family toward the end of the story.[26] In a way, for Danielewski,

American mythology, likewise, seems impenetrable. Despite highlighting the fractured foundations, unrealizable ideals, and the violent past that the nation was built upon, the »darkness sweep[ing] in like a hand« (2000: 528), these myths cannot simply be forgotten or done away with, as they are in continuous circulation, constantly being fabricated and re-fabricated and effortlessly moving from one medium to the next. As Michel Foucault (1962) argued, transgression or subversion can always merely be a gesture that inadvertently reaffirms that which one tries to transgress or subvert. Yet this situation presents not some sort of unsolvable catch-22 Danielewski attempts to fathom, because, in a Whitmanian twist, he embraces this apparent dilemma in *House of Leaves*, suggesting that ›this book, like the United States, contains multitudes.‹ After all, when Karen in the end saves Navy from the house (which simply dissolves into nothingness), and they even marry shortly after, Danielewski emphasizes the values of family life.[27] Thereby, *House of Leaves* eventually reaffirms the values of the ideals, myths, and traditions that hold ›America‹ together. However, as the book truly embraces an ambiguous stance toward these myths, this apparent reaffirmation of core American values isn't as straightforward as it first seems, for repression emerges as one of the—if not *the*—prime agents in the re-creation of the American family through the Navidsons' attempts to forget those terrifying months in Virginia. The symbolic equation of the nuclear family and the American ›family‹, that is, the American nation, thus effectively highlights repression's important role in the construction of the imagined community that is the people of the United States.

Notes

1 | Since this is the first quote taken from *House of Leaves* and although no ›[sic]‹ is needed here, I should stress that I will refrain from using ›[sic]‹ to indicate mistakes in the book—all quotations have been correctly copied from the book, so, any typos, punctuation errors, etc. that are found in the quotations are in the original text. I should also apologize for the somewhat excessive use of notes, which I blame the structure of the book under discussion for.

2 | While Johnny explicitly states that »Zampanò's entire project is about a film which doesn't even exist« (Danielewski 2000: xix) in his introduction to the book, the editors acknowledge the existence of one of the tapes in their credits, expressing »[s]pecial thanks to the Talmor Zedactur Depositary for providing a VHS copy of ›Exploration #4‹« (2000: 708), and they even include a shot taken from »Exploration #4« in their appendix. Truant also somewhat doubts Zampanò's existence other than as a *nom de plume*—that is, an existence as a verbal construct based on a verbal construct (»I never came across any [...] official document insinuating that yes, he indeed was An-Actual-&-Accounted-For person« [2000: xii], he writes).

3 | Admittedly, ›Navidson‹ is a misnomer, for Will and Karen don't get married until after the events in Virginia. However, considering that the family is depicted as rather patriarchal (and for the sake of simplicity), I have, like Zampanò, opted to refer to the family as the ›Navidson‹ rather than the ›Navidson/Green family‹.

4 | The end of chapter seventeen highlights that numerous »people dwelling on *The Navidson Record* have shown an increase in obsessiveness, insomnia, and incoherence: ›Most of those who chose to abandon their interest soon recovered. A few, however, required counseling and in some instances medication and hospitalization. Three cases resulted in suicide‹« (Danielewski 2000: 407).

5 | I will be following in Nele Bemong's footsteps here, for she correctly stresses that »the designation ›novel‹« doesn't seem »to be adequate« (2003: par. 1) in the case of *House of Leaves*.

6 | The fonts used in the illustration mirror the fonts used by the various narratorial voices in *House of Leaves*. From Zampanò's descriptions, it seems as if *The Navidson Record* consists of a series of videos(-within-the-video), which is why one could add yet another level there. Similarly, Pelafina's letters could be considered embedded within Johnny's narrative. Furthermore, real readers are explicitly invited to annotate their copies of the book (»You might try scribbling [...] in the margins of this book« [Danielewski 2000: xxiii]), playfully turning them into parts of the narrative and highlighting that »[p]ersonal experience is at the heart of Danielewski's book« (Brick 2004: par. 8). Finally, the narrative layers not only influence one another (as evidenced by the strong ties between Truant's narrative and Zampanò's manuscript), but, in some cases, the borderlines separating the various diegetic realities are metaleptically transgressed, such as when *House of Leaves* physically (yet verbally constructed) appears in *The Navidson Record* or in Johnny's narrative (Danielewski 2000: 465; 513) before it has actually been completed (for more on metalepsis, see Jeff Thoss' contribution to this volume). However, since this chapter does not revolve around *House of Leaves'* structural complexities, I have decided to construct the diagram in a more simple way to provide an illustrative overview of the book.

7 | Laura Barratt's more recent investigation of the uncanny in the context of *House of Leaves* proves more insightful, as she concentrates on the interrelations between the uncanny and representation, for the »familiarity and strangeness, clarity and obfuscation« of mediation in all its shapes and forms »literally haunt us« (2011: 249).

8 | As does the cover layout of certain editions of *House of Leaves* that tries to visually capture the »goddamn spatial rape« [Danielewski 2000: 55] described in *The Navidson Record*.

9 | On *House of Leaves* and the physical reality of the book, see Starre (2011).

10 | My translation of »l'idée qu'elle est la réalisation de tout ce dont les autres ont rêvé—justice, abondance, droit, richesse, liberté.«

11 | The term is taken from Slavoj Žižek. Even though his take on interpassivity (1997/2008: 144-152) proves to be more complex, he essentially highlights an issue that numerous new media scholars have more recently discussed, namely that there

is no ›truly‹ interactive experience, for whether one talks about a video game or another medium, there are certain rules and boundaries that restrict the movements etc. of the user/recipient. In the case of a book such as *House of Leaves*, these restrictions are provided by the textual cues in the book to which the reader rather reacts than actively acts, among others.

12 | *House of Leaves* scholarship tends to be more attracted to the book's formal and structural features than its narrative and the traditions that the story taps into. N. Katherine Hayles, for example, writes that *House of Leaves* is »camouflaged as a haunted house tale« and »worlds away« from your typical ghost story, since it rather »instantiates the crisis characteristic of post-modernism, in which representation is short-circuited by the realization that there is no reality independent of mediation« (2002: 110). As a result, the members of the Navidson family »are evacuated as originary objects of representation but reconstituted through multiple layers of mediation« (2002: 114). This is a fine argument, indeed, but doesn't change the fact that the cultural echoes of the American Gothic resonate in the book.

13 | Andrew Ross, supposedly literature professor at Princeton, who is interviewed by Karen Green about the (non-existing) monster in the Navidson home, says: »Quite a few Brits you know still prefer their ghosts decked in crepe and cobweb, candelabra in one hand. Your monster, however, is purely American. Edgeless for one thing, something of a compendium of diverse cultures definitely requires. You can't identify this creature with any one group. Its individuality is imperceptible, and like the dark side of the moon, invisible but not without influence« (Danielewski 2000: 357).

14 | As the editors highlight, the photo »is clearly based on Kevin Carter's 1994 Pulitzer Prize-winning photograph of a vulture preying on a tiny Sudanese girl who collapsed on her way to a feeding center« (Danielewski 2000: 368). Kevin Carter went on to commit suicide shortly after winning the Pulitzer, among others because of being »haunted by the vivid memories [...] of starving or wounded children« (qtd. in MacLeod 1994: par. 28).

15 | Even though this connection to Virginian history is not established in the book, Zampanó does quote this bible verse at one point (Danielewski 2000: 249).

16 | History sometimes provides the most excellent examples of irony, for the ›New World‹ was as ›unconquered‹ as the ›virgin‹ queen back home in the ›Old World‹ whose name the new colony was to bear.

17 | As Leslie Fielder has observed, in contrast to the European Gothic, in the American Gothic »the heathen, unredeemed wilderness and not the decaying monuments of a dying class, nature and not society becomes the symbol for evil« (1960/1997: 160).

18 | I don't mean to go into details about the interrelations between the uncanny and the home (as expressed in the German word ›unheimlich‹), for they are repeatedly indicated in *House of Leaves*. For example, Zampanó writes: »[T]hat which is uncanny or unheimlich is neither homey nor protective, nor comforting nor familiar. It is alien, exposed, and unsettling, or in other words, the perfect description of the house on Ash Tree Lane« (Danielewski 2000: 28). And Harold Bloom (yes, the literary critic) notes in an interview that »this ›unhomely‹ might as well be called ›the homely‹, [Freud] observes, ›for this uncanny is in reality nothing new or foreign, but something familiar and old-established in the mind that has been estranged only by the process of repression‹« (2000: 359). But these are just two examples.

19 | It may be highlighted that also other national gothic literatures are often located in the family. The mother of all English gothic texts, *The Castle of Otranto* (1764), after all, tells the story of an ancient curse that befalls the lords of Otranto whenever the family becomes too powerful. The difference is, however, that European gothic tales center on upper-class families, whereas the American Gothic is generally located in the middle class.

20 | It is noteworthy that the escape from the house is first referred to as ›The Evacuation‹ before being titled ›The Escape‹ (Danielewski 2000: 316; 339). This aspect is significant, for nearly all other parts of the video, no matter how important, don't have specific section markers in the text. In light of the note highlighting (possible) alternative paths, *House of Leaves* emphasizes how ›The Escape‹ is merely a construction within another construction, thus underscoring the significant role taken by the element of fabrication in the entire book.

21 | As is the case with more or less any statement about the events represented in *House of Leaves*, one cannot say with certainty that Karen was, in fact, sexually abused. Her older sister Linda talks in a talk show about how their stepfather raped both of the sisters when Karen was fourteen. However, »when asked by various reporters to confirm her sister's claim, Karen refused to comment« (Danielewski 2000: 347). She also »never mentions any history of sexual abuse« (2000: 347) during her therapy against claustrophobia.

22 | »Frostbite claimed his right hand and clipped the top portion of one ear. Patches of skin on his face were also removed as well as his left eye. Furthermore his hap had inexplicably shattered and had to be replaced. Doctors said he would need a crutch for the rest of his life« (Danielewski 2000: 523).

23 | If one doesn't accept *House of Leaves* as a deconstructionist parable about absence, that is. As Will Slocombe writes in his nihilist/deconstructionist reading of *House of Leaves*: »The House, both as house and as text, seeks to unwrite its own creation [...]. [T]his reflexive destruction of its own axioms demonstrates an important aspect of nihilism and the primary difference between nihilism and deconstruction [...]. Although deconstruction attempts to invert and displace the hierarchy of a text, replacing the dominant term with the residual and then shifting the entire system, it accepts that one can never escape from metaphysical underpinnings, that one can only realize what they are. Nihilism, in contrast, can never allow itself to be written and must disappear as it is written: rather than exert a dominant metaphysic that must be undone, nihilism (and nothingness) must never be stable enough to be written about in any secure way« (2005: 92-93). Readers »desire to read ›something‹ into the House because [they] cannot tolerate the absence that it signifies [...]. [I]t is a process of interpretation by which we seek to bring this House into Being. The House continually resists such readings« (2005: 95-96). Consequently, *House of Leaves* »has nothing to say« (2005: 105).

Indeed, with each revision of the chapter at hand, I increasingly felt like I merely found what I was looking for.

24 | On the interconnections between Hollywood and the American Dream, see, for example, Sternheimer (2011).

25 | The circular structure is hard to miss here, for the final images of *The Navidson Record* show a Halloween parade »pass[ing] from sight« before Navidson »focuses on the empty road beyond, a pale curve vanishing into the woods where nothing moves and a street lamp flickers on and off until at last it flickers out and darkness sweeps in like a hand« (Danielewski 2000: 528), as the Navidson narrative concludes with yet another instance of the civilization vs. nature binary. On another level, circularity is essentially about repetition and countering linear conceptions of time, which are key themes not only in Freud's conception of the uncanny, but also Jacques Derrida's (1993) conceptualizaton of hauntology (see Leo Lippert's contribution to this volume for an application of Derrida's concept in a different context).

26 | For a more detailed analysis of the meaning of thresholds in Danielewski's book, see Gibbons (2012: 46-85).

27 | Tellingly, the importance of family life is already highlighted much earlier, for when Navidson first explores the hallway and gets lost, Daisy »tug[s] her daddy home with a cry« (Danielewski 2000: 68).

REFERENCES

Anderson, Benedict (1983/2006): *Imagined Communities: Reflections on the Origin and Spread of Nationalism*, New Edition. London: Verso Books.

Barrett, Laura (2011): »Repetition with a Difference: Representation and the Uncanny in *House of Leaves*«, Horror Studies 2 (2), pp. 247–264.

Barthes, Roland (1970): *S/Z*. Paris: Éditions de Seuil.

Baudrillard, Jean (1986): *Amérique*. Paris: Éditions Grasset & Fasquelle.

Belletto, Steven (2009): »Rescuing Interpretation with Mark Danielewski: The Genre of Scholarship in *House of Leaves*«, Genre: Forms of Discourse and Culture 42 (3–4), pp. 99–117.

Bemong, Nele (2003): »The Uncanny in Mark Z. Danielewski's *House of Leaves*«, Image & Narrative 5, online.

Benesch, Klaus (2005): »Concepts of Space in American Culture: An Introduction«, in Klaus Benesch/Kerstin Schmidt (eds.), *Space in America: Theory—History—Culture*. Amsterdam: Rodopi, pp. 11–21.

Botting, Fred (1996): *Gothic*. London: Routledge.

Brick, Martin (2004): »Blueprint(s): Rubric for a Deconstructed Age in *House of Leaves*«, Philament 2, online.

Chanen, Brian W. (2007): »Surfing the Text: The Digital Environment in Mark Z. Danielewski's *House of Leaves*«, *European Journal of English Studies* 11 (2), pp. 163–176.
Cox, Katharine (2006): »What Has Made Me? Locating Mother in the Textual Labyrinth of Mark Z. Danielewski's *House of Leaves*«, *Critical Survey* 18 (2), pp. 4–15.
Danielewski, Mark Z. (2000): *House of Leaves*. New York: Pantheon Books.
Davenport-Hines, Richard (1998): *Gothic: 400 Years of Excess, Horror, Evil and Ruin*. London: Fourth Estate.
Derrida, Jacques (1993): *Spectres de Marx: L'État de la dette, le travail du deuil et la nouvelle Internationale*. Paris: Éditions Galilée.
Dupuy, Valérie (2008): »Le livre métamorphosé en volume: *La Maison des feuilles* de Mark Z. Danielewski«, *Voix Pluriells* 5 (1), online.
Fielder, Leslie (1960/1997): *Love and Death in the American Novel*, Revised Edition. Champaign: Dalkey Archive.
Foucault, Michel (1962): »Préface à la transgression«, *Critique* 195–196, pp. 751–769.
Freud, Sigmund (1919/1955): »The Uncanny« (trans. James Strachey), in *The Standard Edition of the Complete Psychological Works of Sigmund Freud*, Vol. 17. London: Hogarth Press, pp. 217–256.
Gehring, Melina (2009): »Das Labyrinth als Chronotopos: Raumtheoretische Überlegungen zu Mark Z. Danielewskis *House of Leaves*«, in Wolfgang Hallet/Birgit Neumann (eds.), *Raum und Bewegung in der Literatur: Die Literaturwissenschaften und der Spatial Turn*. Bielefeld: transcript Verlag, pp. 319–334.
Gibbons, Alison (2010): »The Narrative Worlds and Multimodal Figures of *House of Leaves*: ›—find your own words; I have no more‹«, in Marina Grishakova/Marie-Laure Ryan (eds.), *Intermediality and Storytelling*. Berlin: Walter de Gruyter, pp. 285–311.
——— (2012): *Multimodality, Cognition, and Experimental Literature*. London: Routledge.
Goddu, Teresa A. (1997): *Gothic America: Narrative, History and Nation*. New York: Columbia University Press.
Graulund, Rune (2006): »Text and Paratext in Mark Z. Danielewski's *House of Leaves*«, *Word & Image* 22 (4), pp. 379–389.
Hamilton, Natalie (2008): »The A-Mazing House: The Labyrinth as Theme and Form in Mark Z. Danielewski's *House of Leaves*«, *Critique* 50 (1), pp. 3–15.
Hansen, Mark B. N. (2004): »The Digital Topography of Mark Z. Danielewski's *House of Leaves*«, *Contemporary Literature* 45 (4), pp. 597–636.
Hayles, N. Katherine (2002): *Writing Machines*. Cambridge: MIT Press.
Idiart, Jeanette/Jennifer Schultz (1999): »American Gothic Landscapes: The New World to Vietnam«, in Glennis Byron/David Punter (eds.), *Spectral*

Readings: Towards a Gothic Geography. Basingstoke: Palgrave Macmillan, pp. 127–139.

James, Henry (1865/1984): »Miss Braddon«, in Leon Edel/Mark Wilson (eds.), *Henry James: Literary Criticism*, Volume One. New York: Library of America, pp. 741–746.

Kelly, Robert (2000): »Home Sweet Hole«, *New York Times* [online], 26 March, http://www.nytimes.com/books/00/03/26/reviews/000326.26kellyt.html. 12 April 2011.

Little, William G. (2007): »Nothing to Write Home About: Impossible Reception in Mark Z. Danielewski's *House of Leaves*«, in Neil Brooks/Josh Toth (eds.), *The Mourning After: Attending the Wake of Postmodernism*. Amsterdam: Rodopi, pp. 169–199.

MacLeod, Scott (1994): »The Life and Death of Kevin Carter«, *Time* [online], 12 September, http://www.time.com/time/magazine/article/0,9171,981431,00.html. 4 March 2012.

Message, Vincent (2009): »Impossible de s'en sortir seul: Fictions labyrintiques et solitude chez Kafka, Borges, Danielewski et Kubrick«, *Amaltea: Revista de mitocritica* 1, pp. 189–201.

Olson, Charles (1947/1997): *Call Me Ishmael*. Baltimore: Johns Hopkins University Press.

Poole, Steven (2000): »Gothic Scholar«, *The Guardian* [online], 15 July, http://www.guardian.co.uk/books/2000/jul/15/fiction.reviews. 12 April 2011.

Pressman, Jessica (2006): »*House of Leaves*: Reading the Networked Novel«, *Studies in American Fiction* 34 (1), pp. 107–128.

Sedgwick, Eve Kosofsky (1980): *The Coherence of Gothic Conventions*. New York: Arno Press.

Slatta, Richard W. (2010): »Making and Unmaking Myths of the American Frontier«, *European Journal of American Culture* 29 (2), pp. 81–92.

Slocombe, Will (2005): »›This is Not For You‹: Nihilism and the House that Jacques Built«, *Modern Fiction Studies* 51 (1), pp. 88–109.

Slotkin, Richard (1985): *The Fatal Environment: The Myth of the Frontier in the Age of Industrialization, 1800–1890*. Norman: University of Oklahoma Press.

Sternheimer, Karen (2011): *Celebrity Culture and the American Dream: Stardom and Social Mobility*. New York: Routledge.

Starre, Alexander (2011): »The Materiality of Books and TV: *House of Leaves* and *The Sopranos* in a World of Formless Content and Media Competition«, in Katharina Bantleon/Jeff Thoss/Werner Wolf (eds.), *The Metareferential Turn in Contemporary Arts and Media: Forms, Functions, Attempts at Explanation*. Amsterdam: Rodopi, pp. 195–215.

Turner, Frederick Jackson (1920/1986): *The Frontier in American History*. Tuscon: University of Arizona Press.

Vidler, Anthony (1992): *The Architectural Uncanny: Essays in the Modern Unhomely*. Cambridge: MIT Press.
Wheatley, Helen (2006): *Gothic Television*. Manchester: Manchester University Press.
Williams, Tony (1996): *Hearths of Darkness: The Family in the American Horror Film*. Hoboken: Fairleigh Dickinson University Press.
Žižek, Slavoj (1997/2008): *The Plague of Fantasies*. London: Verso Books.
Zolles, Christian (2011): »Mark Z. Danielewskis Roman *House of Leaves* (2000) und das Grauen der Zwischenräume im Zeichen dokumentarischer Medien in Zeiten der Kulturwissenschaften«, in Jörg van Bebber (ed.), *Dawn of an Evil Millennium: Horror/Kultur im neuen Jahrtausend*. Darmstadt: Büchner-Verlag, pp. 27–32.

Drawing Borders

Meeting at the Border
The Canadian ›Two Solitudes‹ in *Bon Cop, Bad Cop*

YVONNE VÖLKL

Canada has two so-called ›Founding Nations‹: Great Britain and France.[1] The French colony of *La Nouvelle France* (New France) was established by the Frenchman Jacques Cartier, who took possession of the land around the Saint Lawrence River in 1534. This French-speaking settlement became a British colony following the French and Indian War (1754–1763), but was at the same time able to remain a Roman Catholic province and maintain French civil law within its borders. This allowed the francophone population in this region to keep their language, religion, and tradition alive throughout the subsequent centuries. Due to the linguistic, religious, and cultural differences, the relationship between French and English Canadians has remained distant and infrequent to this day. No wonder that the ways »the two groups envisage their predicament, their problems, and their common country are so different that it is hard to find a common language. They are like two photographs of the same object taken from such different points of view that they cannot be superimposed« (Taylor 1993: 24). For this reason, one of the most frequently used metaphors to describe the relations between the two communities is the image of the ›two solitudes‹.

Literature, film, and other media have repeatedly employed this image and contributed to keep the ›two solitudes‹ in the public consciousness. One particular example is the Canadian blockbuster *Bon Cop, Bad Cop*. Hence, the present contribution investigates the particular relationship between English and French Canadians as represented in the 2006 motion picture *Bon Cop, Bad Cop* by Canadian filmmaker Érik Canuel. In this movie, the encounters between the two solitudes are repeatedly accompanied by points of friction and affronts, which are due to the borders that these two linguistic communities have drawn in the past. Therefore I will, in a first step, describe the origins of the metaphorical designation ›two solitudes‹. Then, I will briefly elucidate on the question of borders in academic discourse, focusing on how borders are created and how their existence affects the lives of people within such demarcation lines. Finally, I will turn to the movie and examine the representation of anglophone and francophone characters. In the process, I will demonstrate how its comic approach allows for the construction and deconstruction of political, cultural, and linguistic borders.

THE CANADIAN TWO SOLITUDES

The expression ›two solitudes‹ was coined by Hugh MacLennan, an English Canadian author from Nova Scotia, in his famous 1945 novel of the same title. The plot of *Two Solitudes* unfolds in Quebec between the First and Second World War and presents the story of two young lovers who originate from separate communities: Paul comes from a Franco-Catholic family and Heather from an Anglo-Protestant one. When Paul and Heather fall in love, both families are against their relationship. In fact, their distinct background increases the surrounding's prejudices against the respective other. As a result, when the couple wants to marry, Paul and Heather have to face »the legend of a whole race« (MacLennan 1945/2003: 399). In the end, however, Heather and Paul are not victims of their cultural origins, but are able to surmount their cultural heritage and get married. By using the genre of romance, MacLennan renders a detailed portrayal of the French and English Canadian societies, which are aware of each other but live apart within one nation. Thus, the conjugal bond of the protagonists represents a symbolic reconciliation of a Canada that battles the predominance of these conservative legends.

Throughout the twentieth century, the tensions between the two Founding Nations did not disappear. From the 1960s onwards, the francophone community has increasingly tried to further underline its French cultural heritage. Due to the Quiet Revolution, the French-speaking men and women no longer perceive themselves as ›French Canadians‹ (that is, Canadians speaking French) but rather as ›Quebecois‹ (that is, the French-speaking community of Quebec). In 1977, Quebec's politicians introduced the Charter of the French Language, which made, as the preamble reads, »French the language of Government and the Law, as well as the normal and everyday language of work, instruction, communication, commerce and business.« Following this legal move, commonly referred to as ›Bill 101‹, Quebec intensified its individualization process and attempted to separate from the rest of Canada with two referenda in 1980 and in 1995. Those efforts were, however, unsuccessful.

In the twenty-first century, Canadians still face the same tensions between the French- and English-speaking areas as depicted in MacLennan's *Two Solitudes*. The struggle for sovereignty has not come to an end, but even without Quebec's political separation from Canada, there is a host of indicators showing that Quebec belongs less and less to the rest of Canada. This increasing distance between francophone and anglophone Canada is not a one-way street, however, as ›Canada‹ increasingly makes its decisions without Quebec (cf. Dutrisac 2012: par. 4). As a result, a conference on *The Quebec Question for the Next Generation* with the aim of »bringing together eminent thinkers and key players on Quebec and Canada–Quebec relations, as well as—by way of bridging the generations—emerging stars and future players on these matters to frame the ›Quebec Question‹ for this early new century« (Gagnon/Studin/Tuohy 2012: par. 3) was held in February 2012. Apart from the widening political gap, the Island of Montreal, for example, demonstrates that a residential segregation between the two linguistic communities is also very present today. When looking at data from the 2006 Census (cf. Statistics Canada 2009), one can easily see that while anglophone neighborhoods are more likely to be found in the West of the Island, the francophone population[2] rather settles in the East. However, the dividing line between East and West is less clear than in the middle of the twentieth century, when the Saint Lawrence Boulevard constituted a clear-cut borderline between francophones and anglophones.

In the teasers for *Bon Cop, Bad Cop* (both English and French), the linguistic subject matter is stressed in order to depict the two linguistic communities and their (missing) relationship. Yet, there are considerable differences between the two trailers: The French one uses English and French to an equal amount and even suggests communication between the two solitudes through the tagline ›For once, the two solitudes will speak to each other... maybe.‹[3] The English trailer, however, uses »only one, clearly facetious, line of subtitled French dialogue,« thus implying »that the film would be about French Canadians without actually being in *French*« (Macdougall 2011: par. 12). The tagline here, ›Shoot first, translate later‹, points to the linguistic and cultural borders between the two solitudes, which lead to situations of misunderstanding and misinterpretation due to each other's ignorance.

BORDERS

The Oxford Dictionary defines a border as »a line separating two countries, administrative divisions, or other areas« or as »the edge or boundary of something.« The expression commonly refers to a political border that divides two or more nation states from one another and came into being in Europe during the eighteenth century when the modern state system became increasingly dominant. Thus, the development of borders can be explained as »spatial and temporal records of relationships between local communities and between states« (Wilson/Donnan 1998: 5). The people who live within such national boundaries develop a feeling of belonging to their country and start to identify with it. Ian Angus calls this socially defined form of human organization ›social identity‹ (1997). Through the establishment of borders to delimit a national territory, simultaneously a border between the self and the Other is created. Such an outward demarcation usually entails the preservation of one's own, and requires the acceptance of the Other. The border essentializes the identity of a group because the border is imposed on it. Crossing borders then allows each group member to know who and, moreover, where s/he is.

Anthropologists habitually see cultural borders to co-exist with national borders (cf. Erickson 2004: 41; Lugo 1997). Most notably, when national languages of neighboring countries differ, their cultural habits vary likewise. The border therefore assumes an »image of cultural juxtaposition« (Wilson/Donnan 1998: 6). With regard to Canada, it can be observed that Quebec's territory appears equivalent to French heritage, the Catholic Church, and, in particular, the French language, while the rest of Canada is usually related to Anglo-Saxon culture, the Protestant Church, and the English language. Crossing the provincial border to Quebec is, for an English Canadian, not just a physical movement, but a move toward another world. By crossing a border, one can observe how the inhabitants on the other side »shift in language, social relationships, and negotiation of ways to survive while also insisting on maintaining a rich life that does not lose identity and cultural connection, yet provides access to the mainstream world within the region« (Van Dongen 2004: 3–4). When entering the province of Quebec, one realizes that the negotiation of French language rights—within the province as the only official language and in Canada as one of the two official languages next to English—constitutes one of the most significant and striking achievements of the Quebecois. In fact, by preserving their language throughout the past centuries, they have been able to preserve their cultural heritage and identity as well as a sense of belonging. The threat of language loss is a concern which anglophone Canadians never had and therefore can hardly understand.

Whenever people cross borders and venture into the other's territory, their image of this Other is usually rife with stereotypical believes and prejudices. Typically, all members of the observed group are lumped together, which makes the Other appear as a homogenous group, a phenomenon James L. Hilton and William von Hippel refer to as ›homogeneity effect‹ (1996). The application of certain clichés and images, whether true or false, helps the observing group preserve its own system of values. Stereotyping additionally acts »to simplify communication at the social level, allowing people to enjoy an economy of words when speaking about and to others« (Stangor/Schaller 2000: 74). In this sense, stereotypes contain certain pieces of information conveyed by simply evoking the stereotype. For example, in the United States, Latinos are known for their family values and are considered hard-working people, while African Americans are said to ›naturally‹ excel in sports and have ›innate‹ musical talent. Thus, if one talks about Latinos or African Americans in the U.S., the supplementary information does not need to be evoked or explained explicitly, because it is already part of the collective image of that group.

Strictly speaking, stereotypes are timeless and stable structures that facilitate »the transmission of ideas, images, and concepts« (Rosello 1998: 23). Due to their memorability and timelessness, they are branded in people's minds and »start an almost autonomous life as a repeatable unit of ideology« (Rosello 1998: 35). Notwithstanding, »stereotypes are social inventions« which »can be taken apart« (New 1998: 45–46). One way of inverting and subverting these stereotypes and, in a further step, changing attitudes is through the employment of comedy:

Comedy tends to involve departures of a particular kind—or particular kinds—from what are considered to be the ›normal‹ routines of life of the social group in question. In order to be marked out as comic, the events represented—or the mode of representation—tend to be different in characteristic ways from what is usually expected in the non-comic world. Comedy often lies in the gap between the two, [...] including incongruity and exaggeration. (King 2002: 5)

Apart from the potential of being »subversive, [thus] questioning the norms from which it departs,« a comedy can also be »affirmative, [hence] reconfirming what it recognises through the act of departure; or a mixture of the two« (King 2002: 8). Comic stereotypes are, for example, often employed in movies and elsewhere to represent the other, be it other cultures, nations, or customs. It is, indeed, quite »common for one culture to find the norms of another to be foolish or ›unclean‹ in one way or another, a potential source of comedy that helps to mark the bounds of the former« (King 2002: 144). In fact, making fun of dissimilar communities and laughing at them is a common way to define and differentiate one's own group from other collective units. As will be shown in the following passages, the comic representation of the anglophone and francophone characters in *Bon Cop, Bad Cop* does not only underpin dominant stereotypes the two solitudes have of the respective other, but it embodies at the same time a source of criticism of these prejudices.

Bon Cop, Bad Cop: A Brief Introduction

Bon Cop, Bad Cop can be considered a milestone in Canadian cultural history, because it was the first Canadian feature film produced in both official Canadian languages, English and French. There is an English and a French version of the movie with English or French subtitles and even the commentaries to the film are produced in both languages. The purpose behind this strategy was to create

Illustration 1: The differences between not only the two cops but the two cultures they represent are already made explicit on Bon Cop, Bad Cop's *DVD Cover.*

Source: *Bon Cop, Bad Cop* Francophone DVD Cover. *Bon Cop, Bad Cop* © Alliance Atlantis Vivafilm, 2006.

an ›überCanadian‹ action comedy about the two solitudes, as stated by producer Kevin Thierney:

People were going to speak both languages and the jokes were going to be about Canada and Canadian culture, and this lunacy of living side by side. We wanted to see if we could make French people and English people laugh at the same situation. Maybe not exactly the joke, but the same basic situation, to laugh at ourselves and each other. (qtd. in Bokser 2007: par. 11)

The concept proved a success: In 2006, *Bon Cop, Bad Cop* was awarded the Golden Reel Award, since it was the Canadian film with the biggest box office gross of the year.[4] It received the Genie Award for best sound and for best Canadian motion picture. Moreover, the film was awarded the Canadian Comedy Award and won the Jutra Award for best international motion picture in 2008.

The movie begins with the murder of a man whose corpse is found perched on top of the welcome/goodbye sign at the Quebec–Ontario border (see *Illustration 2*). A detective from Quebec, David Bouchard, and a detective from Ontario, Martin Ward, meet for the first time at this crime scene, where the question of jurisdiction arises—who should handle the border-crossing homicide? From this first meeting onwards, the two policemen do not get along at all. Ward from the Toronto Police Department and Bouchard from the Montreal Police Department could not be more different characters. The former speaks English and is a conservative who goes by the book, the latter speaks French and works with unorthodox methods. Eventually, they are forced to collaborate on the investigation the geographical jurisdiction of which is as blurred as the crime's motive. As it turns out, the reason for the multiple homicides is connected to a sport that is popular in all of Canada: hockey. The murderer takes his personal revenge for sending and selling Canadian hockey players to the United States. The differences between Bouchard and Ward put their inquiry at risk. However, as soon as they become personally involved in the investigation (attempted murder of Ward, abduction of Bouchard's daughter), their dissimilarity helps them solve the mystery.

THE POLITICAL BORDER IN *BON COP, BAD COP*

The political border shown in the feature film is the provincial border between Quebec and Ontario. This border is already alluded to in the poster and on the cover of the DVD (see *Illustration 1*). On this picture, one can distinguish one half of the fleur-de-lis, the iris flower, a symbol for French-speaking Quebec pictured four times in its provincial flag, and next to the iris, one half of the maple leaf, a symbol for English-speaking Canada displayed in the middle of the Canadian ensign.

When the two main characters meet for the very first time at the provincial frontier, the only indication of the border is a highway sign. First, Martin Ward holds his police badge, which he carries in his leather wallet, toward Bouchard, but does not transgress the imaginary prolongation of the welcome/goodbye sign. Bouchard examines the badge, but then shows his badge to the detective from Ontario, also decidedly staying within his borders. When turning to the victim on the road sign, Bouchard is quickly convinced that this case is not subject to Quebec jurisdiction, because the feet of the victim are on the Ontario side. Interestingly enough, he asserts that in sports like football and tennis, when a player steps over the line, s/he is out. Conversely, Ward, who, of course, believes the case to be subjected to Quebec jurisdiction because the upper part of the

Illustration 2: The highway sign creates not only a visual borderline between Ontario (left) and Quebec (right).

Source: *Bon Cop, Bad Cop* Blu-Ray. *Bon Cop, Bad Cop* © Alliance Atlantis Vivafilm, 2006.

victim's body is on the Quebec side, argues with the 100-yard dash, where the head and chest break the tape, and horse racing, where the horse wins by a nose.

During this entire scene, the detectives clearly keep standing on their respective side of the border, as if the extension of the highway sign stopped them from crossing the borderline (see *Illustration 2*). It is only after the sports argument that Bouchard traverses this imaginary border to see the victim from the Ontario side. Ward then leads him back to Quebec, whereupon Bouchard reacts with a snappy comment: »C'est quoi là?⁵ Do I need a passport?« Not reacting to Bouchard's questions, Ward (still standing in Ontario) indicates that the suspect must have been a true Quebecois because »his heart is in Quebec.« After Bouchard's waspish reply, »His ass belongs to you,« Ward is tired of the preposterous reasoning. He agrees to take over the case and climbs the road sign with the help of a ladder from Ontario. Bouchard, stubbornly, does not put up with this arrogant treatment and climbs the road sign from the Quebec side. Suddenly both lose their footing, the ladders fall, and the detectives hang onto the victim. The viewers can hear a crack, the body breaks in two halves, and the policemen fall to the ground with the respective body parts—end of scene.

The political border between Quebec and Ontario portrayed in this first meeting between Bouchard and Ward figures as a site of what Gloria Anzaldúa in her conceptualization of borderlands referred to as »open wound« (1987: 3). The unhealed trauma of the French defeat against Great Britain in the Battle of the Plains of Abraham in 1759 and the deep bleeding cut that the two linguistic communities have built since then becomes noticeable through the tense and strained atmosphere during the entire scene. The victim decidedly placed on the border sign embodies this wound. The breaking of the body into two halves indicates that the wounds of the past will come to haunt the entire movie. On the other hand, though, the body represents Canada as a whole, which transgresses the provincial boundaries between Quebec and Ontario—a concept that transcends the more earthly notions of Quebecois and ›Canadian‹. Thus, the corpse in this early scene already hints at the deconstruction of the borders that is characteristic of the entire movie.

THE CULTURAL BORDER IN *BON COP, BAD COP*

Throughout the film, viewers are confronted with stereotypical portrayals of the two solitudes. The anglophone character is depicted as »the uptight, turtleneck-wearing bilingual Ontario law enforcer,« whereas the francophone character is shown as the »explosive, chain-smoking Québecois [sic] cop who cheerfully breaks the rules and drives a jalopy« (Stojanova 2011: par. 2).

Right from the start, the appearance of the policeman from Quebec is disheveled, unshaven, and tousled. He usually wears jeans and a shirt with a checkered shirt or a leather jacket over it (see *Illustration 1*). His manner is insolent, casual, and rebellious. Ward describes him as a »Rambo on steroids« due to his libidinal behavior in scenes such as the breaking into a suspect's house, because he is unwilling to wait for a warrant. Bouchard's arrival at the opening crime scene is shot from an oblique low angle and is accompanied by rock music. Framing and sound underscore the rough and rebellious character of the detective from Montreal. When closing the door of his car, though, the rear-view mirror falls to the ground, the music stops abruptly, and the camera resets to its straight-on angle. Through the breaking of the mirror, the established image of David Bouchard as a rough guy suddenly loses credibility and adds a comic touch. Hereafter, softer music sets in and Martin Ward approaches Bouchard.

Ward's clothing, by contrast—he usually wears a suit (see *Illustration 1*)—suggests that his conduct is more straight-lined and his character stiffer. As it instantly turns out, his investigative methods are more elaborate, since he always has a small digital camera with him and takes pictures of the victims first. He never touches any evidence (for instance, he picks up the hockey mask that was worn by the suspect with a pencil) and repeatedly reminds Bouchard to refrain from touching anything. Moreover, at the end of the film, when the two detectives are about to meet the murderer, the careful Ward takes a bullet-proof vest and hands one over to Bouchard, who refuses to wear it.

Altogether, the depiction of the two main characters wallows in stereotypes. The anglophone character is portrayed as stiff, sophisticated, and polite, whereas the francophone character is ascribed with impoliteness, roughness, and dirt. A similar image already prevailed in the early 1970s when Frances E. Aboud and Donald M. Taylor investigated the respective views of French and English Canadians. At that time, English-speaking Canadians were usually described as intelligent, logical, dominant, and strong. The image of French-speaking Canadians, by contrast, was being affectionate, talkative, artistic, and sensitive (1971: 23).

When thinking of detective Martin Ward, the stereotypical depiction of anglophones as discussed by Aboud and Taylor still holds true—even forty years later. As for the francophone stereotype, it can be detected that Bouchard is definitely very talkative, as he is continuously making jokes about his colleagues, Martin Ward, and the victims. He always has a witty word on his lips that alienates his partner, embarrasses his ex-wife, or attracts Ward's sister. Bouchard's character is certainly also sensitive, because he still seems to have feelings for his ex-wife and feels offended when Ward shows interest in her. However, the affectionate and artistic ascriptions regarding francophones do not really apply to Bouchard except for a scene in which he sheds tears when confronting his ex-wife with the abduction of their daughter.

In general, the expression of emotions is more distinctive on the francophone side. David Bouchard is usually the one who acts before he thinks and is much quicker to show his feelings. Furthermore, Captain LeBœuf from Quebec proves to be a rather choleric character when Bouchard shows no interest in working with the officer from Ontario. Conversely, the anglophone characters, Captain MacDuff and Detective Ward, avoid acting out their feelings in public when they find themselves in the same situation. Captain LeBœuf also acts franticly when the two detectives seem to have muddled the case in the middle of the film, whereas Captain MacDuff raises his voice on the phone only a little in order to emphasize that Ward should return immediately to Toronto, where they would talk about the incident in person.

The only anglophone character with an outgoing personality in the film is Ward's sister Iris, who does not care so much about preserving appearances as her male anglophone counterparts do. She blatantly flirts with Bouchard at the dinner table and takes him to her apartment, where they enjoy each other's brute lust. The following »sexual victory« by the francophone over the anglophone is »internally deconstructed, deceiving and misplacing the expectations, associated with stereotypical utterances and situations« (Stojanova 2011: par. 10). At the final climax, Iris, for example, moans Charles de Gaulle's 1967 salute to the people of Quebec: »Vive le Québec libre!«[6] This barefaced denigration of Quebec's patriotic history reinforces the political views of the Quebecois—since it is Bouchard who teaches the sentence to Iris—, yet it also questions the relevance of the separatist movement in contemporary Canada.

Only toward the end of the movie, when the case becomes personal, Martin Ward puts his exemplary manners and rulebook aside and starts adopting Bouchard's rough working behavior. The two detectives kidnap the assassin's next victim, Mr. Buttman, the commissioner of the hockey league, in order to be able to exchange him for Bouchard's daughter. Ward and Bouchard put Buttman in the trunk of their car and Ward explains to the puzzled abductee that this »is a Quebec tradition«. In this scene, Bouchard even becomes more compassionate and indicates to his partner that he could lose his job over what they are going to do. In response, Ward takes up a Quebec swearing expression and answers him in a way he would have never done before: »Je m'en câlis!«[7]

Another stereotypic label evoked in the film is the mythical designation both communities have reserved for one another. English-speaking Canadians often refer to the francophones from Quebec as ›frogs‹, an allusion to the European heritage of the Quebecois. French people from France are known for their *haute cuisine* which regards frog's legs as a delicacy. Although his French heritage would suggest so, Bouchard never tries frog's legs in the movie, nor does he know how to cook a proper meal. The cooking capabilities are attributed to his counterpart, the highly sophisticated anglophone, who invites Bouchard to a homemade family dinner at his house in the second half of the film.

At times, the Quebecois characters refer to anglophone Canadians as ›tête carrée‹, a term that is directly translated from the English word ›square head‹, but is not used to refer to Germans or Nordic peoples, as is often done in the United States (cf. Meney 1999: 1727). This designation emanates from Bouchard's mouth when he meets Ward in the office of his superior Captain LeBœuf, and again when Captain LeBœuf orders Bouchard to bring Ward, the ›tête carrée‹, to the airport.

The forthright pronunciation of these seemingly unimportant expressions along with the stereotypic depiction can be interpreted as a way of crossing the border to the Other in order to overcome preconceptions. The consistently humorous approach »leads to more tolerant attitudes toward minority groups [or other groups in general] as it provides a lighthearted, indirect way of transversing cultural difference« (Martin 2011: 16). By having the characters overtly repeat their respective nicknames and showing each other's cultural stereotypes, the producers of the film give francophone and anglophone spectators an inside view on the discourse about the Other and about the Self; an insight which may lead to a better understanding and, maybe one day, even to a dismantling of the aforementioned stereotypes.

THE LINGUISTIC BORDER IN *BON COP, BAD COP*

Although Canada is a country that is officially bilingual, the official provincial language in Quebec is French. This means that federal institutions are obliged to serve their customers in English and in French, while provincial institutions in Quebec only need to provide their services in French. The language border, consequently, is very real in the country, as it is the legal means of defining the regions in which a particular language is official.

This territorial status of the language is explicitly brought up when Bouchard tells Ward that in Quebec, he has to work in French. The working language comes up again in a bar fight scene, in which the detectives try to arrest the suspect Luc Therrien. First, Ward fights with Therrien and is forced into a tight corner by the physically stronger adversary. Ward calls out for help to Bouchard, but as he is doing this in English, Bouchard pretends not to understand. Only when Ward addresses him in French is he ready to take action. Then, Bouchard overtakes Therrien, but the suspect liberates himself and knocks Bouchard down. On the floor, Bouchard cries for help. Ward now turns the ›language game‹ around and only assists Bouchard with the arrest once he has asked for support in English.

This fight for language rights is a typical feature of Quebec's singularity within the Canadian nation and can be ascribed to the Charter of the French Language, which aims to reduce the influence of English in the daily lives of the Quebecois. Today, for example, visitors of Montreal will see bilingual signposts on which the French letters are bigger than their English equivalents or on which the French designation for an office or facility appears first. The French-speaking population takes its task of preserving its language and culture seriously. Hence, it is not unusual that visitors are confronted with the resentment of a Quebecois who will (although s/he might speak English) not speak to them when first addressed in English.

What visitors to Quebec, fluent in the French language, may also notice in the province—apart from its distinct accent—is its lexical diversity, which is connected to Quebec's history. From its beginnings until the 1960s, Quebec was strongly influenced by the Catholic Church in all social domains. Out of oppression, people started using words and expressions related to Catholicism and its liturgy to express profanity. These swearwords are usually referred to as ›les sacres‹, which literally means the ›consecrated ones‹. Quebecois use these words in their everyday language, and, of course, there are stronger and milder forms of swearwords, which Bouchard explains to Ward at length after the bar fight. He clarifies the nuances of swearwords as well as swearword constructions and underlines to what extent these expressions are common in the Quebecois language. Ward, obviously astonished by such a great variety of profanities, picks up swearing later on in the film, after the explosion of their car. Yet, the outcome is very clumsy: »Shit de merde de shit de fuck de tabernacle.« The profanities reappear when Bouchard finds himself in a reversed bathtub trying to escape a blaze in the basement of a suspect's house as well as when the two police officers are in front of Captain LeBœuf and the latter starts his reprimand with a tirade of swearwords.

Moreover, there is a tremendous amount of code-switching involved in almost every single dialogue between anglophone and francophone characters, which »gives a hybrid perspective to the work. Code-switching can emphasize the bilingual person's experiences of multiplicity, can show authenticity through the vehicle of language, and can point out aspects of multicultural experience« (Martin 2011: 18). The change from one language to another can be very subtle, as is the case during the first encounter at the provincial border where Bouchard addresses Ward mostly in English and even if addressed in French, the other seems to understand. In the following official meeting between Bouchard, Ward, and their supervisors in Captain LeBœuf's office, the language barrier receives much more attention, as the francophone chief interprets Captain MacDuff's words theatrically until Bouchard admits that he can understand English. In the same scene, the difficulties French people have in pronouncing certain English words is alluded to, as well. LeBœuf cannot pronounce the word ›opportunity‹ properly and is thus corrected by MacDuff until Ward declares his French expertise and liberates LeBœuf from the pretentious treatment by the captain from Ontario.

Linguistic similarities are again brought up and made fun of in the autopsy scene: the corpse of the first victim has a tattoo on his right wrist and the francophone pathologist, wanting to use the correct English word for Ward, asks for the English translation of ›tattoo‹. Yet, the only difference lies in the accent: ›tattoo‹ in English shows an emphasis on the ›a‹-sound, while ›tattoo‹ in French has an emphasis on the ›u‹-sound. This scene thus suggests that the differences between francophone and anglophone Canadians are only of a superficial nature.

Conclusion

Bon Cop, Bad Cop successfully illustrates the centrality of the two solitudes to the Canadian identity. In fact, the film suggests that French and English Canadians complement each other, since Bouchard and Ward can only solve the murder case by working together. Yet collaboration is not easy for the representatives of these two distinct cultures. Therefore, in order to act effectively as a team, both have to take a step toward the other, and, at the same time, they have to transgress borders. Owing to the enforced geographical, cultural, and linguistic displacement, the two main characters are eventually able to surmount the ›open wound‹ between English and French Canadians inflicted by the Plains of Abraham.

The exaggerated use of stereotypes in *Bon Cop, Bad Cop* creates a comic effect that usually makes both English and French Canadian viewers laugh a lot throughout the film. While watching the motion picture, anglophone and francophone spectators do not only perceive the stereotypes they have about the respective Other, but their awareness about those existing about themselves is heightened, too. One can argue that the hybrid narrative that allows for the comic representation of both the anglophone and francophone perspective can be interpreted as a hyperbolical tactic with the purpose of deconstructing the political, cultural, and linguistic borders between Quebec and Ontario. In *Bon Cop, Bad Cop*, laughter does not act as a ›means of limitation‹ but as a ›means of transgression‹ (cf. Bachmaier 2005: 123). Laughter here has a cathartic effect, since laughing at oneself and one's deficiencies strengthens the sense of self-esteem and releases energy which has been used for suppressing and hiding these deficiencies (cf. Bachmaier 2005: 133). Moreover, the viewer is urged to rethink his/her own views with regard to the Other.

As a result, the movie promotes the initial meaning of the two solitudes intended by Hugh MacLennan, who included the following poem by Rainer Maria Rilke at the beginning of his novel: »Love consists in this, / that two solitudes protect, / and touch, and greet each other« (1945/2003: V). In this poem, Rilke refers to two individuals who protect, talk to, and complement each other. Thus, originally, the two solitudes were a symbol of union and only over time was their image changed into the solitary parties that are opposed to and do not communicate with each other. *Bon Cop, Bad Cop* evokes this image of the two solitudes, but resets its meaning back to its original connotation.

Notes

1 | Commonly only the English and French empires are considered Canada's Founding Nations, yet it should not be forgotten that the First Nations peoples were established on Canadian soil long before the beginning of the European settlement.

2 | One has to take into consideration that, nowadays, French is not only spoken by Quebecois, but also by a lot of first-, second-, and third-generation immigrants.

3 | »Pour une fois, les deux solitudes vont se parler... peut-être« (All translations are mine).
4 | It has to be noted that about 90% of the Canadian box office income was generated in Quebec (cf. Stojanova 2011: par. 1).
5 | »What's that?«
6 | »Long live a free Quebec!«
7 | »I don't give a shit!«

REFERENCES

Aboud, Frances E./Taylor, Donald M. (1971): »Ethnic and Role Stereotypes: Their Relative Importance in Person Perception«, *The Journal of Social Psychology* 85, pp. 17–27.

Angus, Ian (1997): *A Border Within: National Identity, Cultural Plurality, and Wilderness*. Montreal: McGill-Queen's University Press.

Anzaldúa, Gloria (1987): *Borderlands/La Frontera: The New Mestiza*. San Francisco: Aunt Luke Book Company.

Bachmaier, Helmut (2005): »Nachwort«, in Helmut Bachmaier (ed.), *Texte zur Theorie der Komik*. Stuttgart: Reclam, pp. 121–134.

Bon Cop, Bad Cop (2006) (CAN, dir. Érik Canuel).

Bokser, Howard (2007): »Park Ex Producer«, *Concordia University Magazine* [online], n. d., http://magazine.concordia.ca/2007/winter/features/Park_Ex.shtml. 22 January 2012.

Dutrisac, Robert (2012): »L'avenir du Québec discuté à Toronto«, *Le Devoir* [online], 3 February, http://www.ledevoir.com/politique/canada/341728/l-avenir-du-quebec-discute-a-toronto. 3 February 2012.

Erickson, Frederick (2004): »Culture in Society and in Educational Practices«, in James A. Banks/Cherry A. M. Banks (eds.), *Multicultural Education: Issues and Perspectives*. Hoboken: John Wiley, pp. 31–60.

Gagnon, Alain/Studin, Irvin/Tuohy, Carolyn (2012): »The Quebec Question for the Next Generation«, *The Quebec Question for the Next Generation* [online], n. d., http://quebecquestionconference.ca/?page_id=19. 3 February 2012.

Hilton, James L./Von Hippel, William (1996): »Stereotypes«, *Annual Review of Psychology* 47, pp. 237–271.

King, Geoff (2002): *Film Comedy*. London: Wallflower Press.

Lugo, Alejandro (1997): »Reflections on Border Theory, Culture, and the Nation«, in Scott Michaelsen/David E. Johnson (eds.), *Border Theory: The Limits of Cultural Politics*. Minneapolis: University of Minnesota Press, pp. 43–67.

Macdougall, Heather (2011): »Facing off: French and English in *Bon Cop, Bad Cop*«, *Reconstruction: Studies in Contemporary Culture* 11 (1), online.

MacLennan, Hugh (1945/2003): *Two Solitudes*, Toronto: Collins.

Martin, Holly E. (2011): *Writing between Cultures: A Study of Hybrid Narratives in Ethnic Literature of the United States*. Jefferson: McFarland.

Meney, Lionel (1999): *Dictionnaire québécois français: Mieux se comprendre entre francophones*. Montréal: Guérin.

New, William H. (1998): *Borderlands: How We Talk about Canada*. Vancouver: University of British Columbia Press.

Rosello, Mireille (1998): *Declining the Stereotypes: Ethnicity and Representation in French Cultures*. Hanover: University Press of New England.

Stangor, Charles/Schaller, Mark (2000): »Stereotypes as Individual and Collective Representations«, in Charles Stangor (ed.), *Stereotypes and Prejudice*. Philadelphia: Psychology Press. pp. 64–82.

Statistics Canada (2009): »2006 Census: The Evolving Linguistic Portrait, 2006 Census: Data tables, figures and maps«, *Statistics Canada* [online], n. d., http://www12.statcan.ca/census-recensement/2006/as-sa/97-555/tables-tableaux-notes-eng.cfm. 20 January 2012.

Stojanova, Christina (2008): »The Bon, the Bad and the Others: Some Remarks on the Phenomenal Success of *Bon Cop, Bad Cop* in Quebec«, *Kinema: A Journal for Film and Audiovisual Media* 16 (2), online.

Taylor, Charles (1993): *Reconciling the Solitudes: Essays on Canadian Federalism and Nationalism*. Montreal: McGill-Queen's University Press.

Van Dongen, Richard (2004): »Reading Across Cultural Borders. Indigenous Influences on Diversity in Children's Literature«, *29th IBBY Congress* [online], n. d., www.sacbf.org.za/2004%20papers/Richard%20van%20Dongen.rtf. 16 June 2011.

Wilson, Thomas M./Donnan, Hastings (1998): »Nation, State and Identity at International Borders«, in Thomas M. Wilson/Hastings Donnan (eds.), *Border Identities: Nation and State at International Frontiers*. Cambridge: Cambridge University Press, pp. 1–30.

›Romanized Gauls‹
The Significance of the United States and the Canada-U.S. Border for Canadian National Identity Construction

Evelyn P. Mayer

> We're like Romanized Gauls—look like Romans, dress like Romans, but aren't Romans—peering over the wall at the real Romans.
> MARGARET ATWOOD, *CURIOUS PURSUITS*, 2005

In her »Letter to America«, leading Canadian author Margaret Atwood explicitly links Americans to Romans and Canadians to Gauls in order to highlight Canadians' perceived inferior role in comparison with Americans, citizens of an alleged neo-imperial power. The possibly identity-effacing kinship between Americans and Canadians becomes evident from this comparison, as does the Americanization of Canadians. Americanized Canadians are like »Romanized Gauls[,] [...] peering over the wall at the real Romans« (2005/2006: 326). Canadians' gaze is thus directed across the border to the United States of America, the powerful neighbor to the South in this asymmetric bilateral relationship. Asymmetry in terms of the economy, politics, the military, population, and geography persists as do differing perceptions pertaining to identities and geopolitical roles.

Atwood postulates that »we've never understood you completely, up here north of the 49th parallel« (2005/2006: 326). This lack of comprehension is intensified by the new American security imperative resulting in legislation such as the Western Hemisphere Travel Initiative in the aftermath of 9/11. Benign historical relations between the two neighboring countries are thus overshadowed by the specter of terrorism and unilateral interests in the security-driven decade after the fall of the Twin Towers. As Peter Andreas states, »Marking and maintaining the border had been such a low priority that it had famously been dubbed ›the longest undefended border in the world‹. This openness, which had historically been a source of mutual pride and even celebration, was suddenly a source of high anxiety« (2005: 454). The shift to anxious rebordering and the »thickening« (Ackleson 2009: 336) of the border threatens local cohesion and cross-border communities.

The assessment of security needs markedly differs between the Canadian and American official stances. This is linked to mutual misconceptions and the alleged lack of knowledge on the part of Americans, even though this is nothing new: »The asymmetry in attention and attitude has been presented as a long-standing feature of the bilateral relationship« (Nord 2011: 391). The interest Canadians have in many things American is not shared by Americans. Ironically, in the security era of post-9/11, this lack of interest has been replaced by too much and unfavorable interest. As Andreas argues, »Although Canadians have long complained of being ignored and taken for granted by the United States (treated as if the border did not exist and Canada was the 51st state), after 9/11 they quickly discovered that the only thing worse than no attention is negative attention« (2005: 454). Ostensible equation of Canada with yet another sort of American place in the American imagination leads to Canadians', especially Anglophone Canadians', urge to stress their distinctiveness as non-Americans, hence to actively engage in self-identification.

Throughout much of her career, Atwood has tapped into the aforementioned notions by drawing attention to the discrepancy that might exist between friendly intentions and negative outcomes. In *Curious Pursuits*, she, for example, writes: »When the Jolly Green Giant goes on the rampage, many lesser plants and animals get trampled underfoot« (2005/2006: 327). Here, she stresses the notion of asymmetry, especially in economic terms, by introducing the image of the ›Jolly Green Giant‹, the superpower, in comparison with Canada. A similar view is held by James Laxer, who ascribes a split or »Jekyll and Hyde personality« (2003: 13) to the United States. From the Canadian perspective, American political behavior can often be volatile and contradictory, underlining a special friendship as allied forces on the one hand while at the same time pursuing rebordering on the other. In fact, it emerges that for any consideration of Canadian national identity, the Canada–U.S border takes on an important symbolic significance and is in most debates coalesced with the majority Anglophone cultural identity. Effectively, the primary function of this border is to maintain both a geographic and symbolic border between ›us‹ and ›them‹, even though the two are interconnected, for any notion of ›we‹ would not exist without ›them‹ to begin with.

However, in a multicultural and diverse society such as Canada, any attempt at constructing a national identity—especially in the twenty-first century—is complicated by its internal fissures and heterogeneity. Cynthia Sugars' nuanced approach, for example, underscores that the issue of national identity construction cannot be relegated to a minority group in Canadian society, but is prevalent on a larger scale and consequently merits scholarly attention. She acknowledges the inherently hierarchical and constructed nature of meta-narratives serving the purpose of forming group identifications and concedes that »various asymmetries of power are perpetuated« (Sugars 2006: 121–122). At the same time, Sugars identifies »the continued appeal of popular representations of Canadian national and cultural identity« (2006: 122). A solution to circumvent the danger of »either reject[ing] the appeals of national identity constructs out of hand, or risk[ing] becoming complicit with an outmoded, homogenizing system marked by intolerance and naïveté« (Sugars 2006: 122) is to openly address these challenges and mindfully negotiate one's own subject position.

Bordering and Canadian National Identity Construction

Canadian identity encompasses an understanding of Canada as a Northern, indigenous, multicultural, and diverse country with the Anglo-Canadian versus Franco-Canadian duality at its heart (see Yvonne Völkl's contribution to this volume). However, Canadian identity is also constructed in opposition to another nation, for the foil of Americanness is key in any attempt to construct a unified Canadian national identity. It is an act of conscious delineation, of bordering, particularly by the Anglophone majority. Garry Sherbert postulates that »Canada's ongoing struggle to define its uniqueness demonstrates that identity is relational, meaning that a group's identity is defined by its similarity to and difference from the identity of another group« (2006: 3–4). Identity needs an Other, and thus, regarding Canada and the United States, both countries »are particularly suited for throwing light on each other's differences« (Morrison 2003: 1). There are ostensible similarities, but cultural distinctions and differing perceptions of their own countries, each other, and the world at large persist below the surface.

The dimension of the North presents an important affirmative and arguably affective aspect of Canadian identity as opposed to merely not being American. Alastair Campbell and Kirk Cameron, however, contest this notion by juxtaposing the importance of the North for the Canadian imaginary versus concrete southern pull factors. They contend that the weight of the North is only marginal due to remoteness and lack of population and posit that »[m]ost Canadians live within 350 kilometres of the American border; their outlook is formed within this ›southern‹ geopolitical zone« (2006: 143). Consequently, Canadians' actions and self-identification are shaped by proximity to the Canada–U.S. border and to their southern neighbor. There is a true North–South divide, not only in terms of Canada and the United States, but also within Canada. Northerners feel neglected by their own compatriots in the southern parts of the country and »that their interests are misunderstood, misrepresented, or simply disregarded by decision makers and interest groups in southern Canada« (Campbell/Cameron 2006: 144). Power lies with the southern economies, whether with regard to Canada or to Canada–U.S. bilateral relations, a picture that could change for internal Canadian affairs with more accessible natural resources in the North of Canada. This is an indication of the internal regional divisions in Canada.

Much like its southern neighbor, Canada is almost too diverse and multi-faceted to talk about a singular Canadian identity, since both identity markers, that is, ›Canadian‹ and ›identity‹, are multiple, plural, and elusive. Nonetheless, some overarching themes pertaining to a construction of Canadian identity can be alluded to, especially when compared to the United States. Zoë Druick states that for a construction of a unified Canadian national identity »[f]rom a centralizing, nationalist point of view« (2006: 86) regional, linguistic, and U.S. media influences can complicate matters. She claims, however, that these phenomena are only seen as problematic if having »an ideal notion of what a nation should be: monolingual, monocultural, bounded, and impervious to other nations' cultural products« (2006: 86). Thus, one solution could be to question monolithic and nationalist viewpoints to embrace a multi-faceted and non-centralized notion of Canadian identity. Druick postulates that the Canadian approach is »the assertion of national identity as a progressive, anti-American strategy« (2006: 86). The actual affirmative character of Canadian national identity lies in the composite of multiple regional and ethnic identities. By embracing differences instead of striving for artificial unity, a more unified Canada can be achieved. Nonetheless, Druick sees multiculturalism in a critical light and argues that Canada is, in fact, »[p]aying lip service to multiculturalism« (2006: 89), for certain ethnic or linguistic groups, such as Anglophone and Francophone Canadians, still hold privileged positions among the cultural groups in Canada.

These are, however, just some of the factors that contribute to Canadian national identity remaining elusive, since defining characteristics of nationhood such as shared national narratives or myths are ambiguous in the Canadian imagination. As Robert Kroetsch suggests, »Canadians cannot agree on what their metanarrative is« and postulates that »in some perverse way, this very falling-apart of our story is what holds our story together« (1997: 355). A defining feature of what it means to be Canadian is not being able to define Canadian identity in a satisfactory manner. Canada, in contrast to the United States, lacks a precise »place and moment of origin« (Kroetsch 1997: 360). Kroetsch reinterprets the perceived lack of founding myths and moments and highlights this absence as a powerful and intangible Canadian presence. He exemplifies his argument by focusing on questions related to absence and margins in Atwood's 1972 novel *Surfacing*: »Again, all is periphery and margin, against the hole in the middle. We are held together by that absence. There is no centre. This disunity is our unity« (1997: 363). Once again, the ambiguity and impossibility of definitive identity markers becomes constitutive of Canadian identity or the Canadian national character: »National meanings are produced through a complex interplay of absence and inclusion« (Bociurkiw 2011: 12). In the same vein, Sugars underscores that absence can become inextricably linked with the adjective ›Canadian‹ and argues that borders are of utmost importance in delineating and defining ambiguous cultures such as Canada: »The more that Canada is identified as lacking in culture, the more that absence is identified as characteristically Canadian; the more amorphous the culture, the more it must be fixed within national boundaries« (2006: 124).

Since Canada's differences from its southern neighbor are central to any definition of Canadianness, mental bordering associated with the physical boundary between Canada and the United States proves a key component of Canadian identity construction. Geographical boundary markers thus demarcate not only sovereign territory, but also serve as symbolic markers of difference. The importance of the notion ›difference‹ in identity construction cannot be overestimated, which is why to be erroneously identified as being part of what is usually considered the Other is especially troubling. In Atwood's novel *Surfacing*, the unnamed Canadian narrator mistakes her own countrymen for Americans and is in turn herself perceived as American by them:

»Say, what part of the States are you all from? It's hard to tell, from your accent. Fred and me guessed Ohio.«
»We're not from the States,« I said, annoyed that he'd mistaken me for one of them.
»No kidding?« His face lit up, he'd seen a real native. »You from here?«
»Yes,« I said. »We all are.«
»So are we,« said the back one unexpectedly. [...]
I was furious with them, they'd disguised themselves. (1972/1979: 122)

The narrator rejects being falsely appropriated as American and is very surprised to learn that the presumed Americans are, in fact, Canadians. Russell Brown argues that »[f]or a Canadian, to be mistaken for an American is to have one's sense of cultural identity threatened« (1991: 3), and this threat of losing one's cultural identity is essentially the point in this brief extract from Atwood's novel. Another example succinctly illustrates the confusing and almost threatening potential of mistaken identities. David, one of the central characters in *Surfacing*, tells the other characters about an incident happening in New York:

»Gentlemen,« David said, raising his tin cup, »Up the Queen. Did that once in a bar in New York and these three Limeys came over and wanted to start a fight, they thought we were Yanks insulting their Queen. But I said she was our Queen too so we had the right, and they ended up buying us a drink.« (Atwood 1972/1979: 104)

This brief anecdote underscores the British legacy of Canada and the Commonwealth connection. Canada can be described as »an Atlantic and an American society, [...] so much so that these two borders came to seem *normative* angles of cultural disposition« (New 1998: 6). Canada has always been influenced by both England and the United States. British, Canadian, and American relations are as much entangled as are the respective national identities.

In *Surfacing*, one of the central topics—aside from the narrator's search for »her self [...], her past [...], and her identity (private and national)« (Hutcheon 1988: 144)—is the border theme, both literally and metaphorically. In its literal meaning, together with her friends, the narrator crosses the Canada–U.S. border into Quebec on her quest for her parents as well as for her past and her self: »There's nothing I can remember till we reach the border, marked by the sign that says BIENVENUE on one side and WELCOME on the other« (Atwood 1972/1979: 5). Atwood reflects on spatial components not only in fiction, but also in connection with her personal sense of Canadian identity by underlining the useful nature of writing in order to find one's place or home:

Because where I live is where everyone lives: it isn't just a place or a region, though it is also that [...]. It's a space composed of images, experiences, the weather, your own past and your ancestors', [...]. The images come from the outside, they are there [...]. But the judgements and the connections (what does it *mean*?) have to be made inside your head and they are made with words: good, bad, like, dislike, whether to go, whether to stay, whether to live there any more. For me that's partly what writing is: an exploration of where in reality I live. (Atwood 2005/2006: 11-12)

The author describes her home country Canada as a place Canadians consciously have to decide to continue to reside in, as emigration is such an easy option thanks to cultural and linguistic knowledge of ›related‹ countries: »I think Canada, more than most countries, is a place you choose to live in. It's easy to leave, and many of us have. There's the US and England, we've been taught more about their histories than our own, we can blend in, become permanent tourists« (Atwood 2005/2006: 12). The image of ›permanent tourists‹ evokes notions of belonging and longing and questions the ideas of roots and home. As a tourist, a person is an outsider, but in making this state permanent, the boundaries between insider and outsider become blurred. A visitor who lives in another country permanently is not a visitor any longer, but might not be a full-fledged insider, either. Such a person inhabits an in-between space, a position that can be simultaneously enabling or disabling for a person's agency or status. A related powerful image is ›amputation‹, actively severing a part that naturally belongs. Atwood postulates that to deny one's roots is equivalent to a self-inflicted »act of amputation: you may become free floating, a citizen of the world (and in what other country is that an ambition?) but only at the cost of arms, legs or heart. By discovering your place you discover yourself« (Atwood 2005/2006: 12). For Atwood, there is a nexus between place and identity; identity formation is strongly dependent on locality. This method uses an affirmative approach to identity, often in sharp contrast to the Anglo-Canadian identity construction based on negation. Atwood compares the »much-sought Canadian identity« (2005/2006: 12) with an experience and feeling she had at Stonehenge: »I did get to Stonehenge, incidentally. I felt at home with it. It was pre-rational, and pre-British, and geological. Nobody knew how it had arrived where it was, or why, or why it had continued to exist; but there it sat, challenging gravity, defying analysis. In fact, it was sort of Canadian« (Atwood 2005/2006: 196). Canada's very existence might be elusive, yet the country does exist as a sovereign entity on the North American continent, thus defying absorption into full-blown American dominance and dependency. Canadians are, however, simply too absorbed with preoccupations concerning Americans, as Atwood puts it: »They'd become addicted to the one-way mirror of the Canadian-American border—we can see you, you can't see us—and had neglected that other mirror, their own culture. The States is an escape fantasy for Canadians. Their own culture shows them what they really look like, and that's always a little hard to take« (1982: 385). The United States are thus simultaneously a point of differentiation and an attractive alternative for Canadians who might find fault with themselves. However, this critical self-reflexive stance is not shared by others.

For people from other countries, Canada presents an »anti-environment that renders the United States more acceptable and intelligible to many small countries of the world« (McLuhan 1977: 227). Marshall McLuhan describes Canada as a mediator between the industrialized and developing countries, since Canada is »[s]haring the American way, without commitment to American goals or responsibilities[, which] makes the Canadian intellectually detached and observant as an interpreter of the American destiny« (1977: 227). Canada is the honest broker and alternative model to the United States, similar, yet different enough. Atwood elaborates that Canada has a truly international perspective which is precisely lacking from the imperial United States:

Canada sees itself as part of the world; a small sinking Titanic squashed between two icebergs, perhaps, but still inevitably a part. The States, on the other hand, has always had a little trouble with games like chess. Situational strategy is difficult if all you can see is your own borders, and beyond that some wispy brownish fuzz that is barely worth considering. The Canadian experience was a circumference with no centre, the American one a centre which was mistaken for the whole thing. (Atwood 1982: 379)

Atwood stresses Canada's multilateral approach and aptness at »situational strategy« in contrast to the United States. Her image of Canada as ›a small sinking Titanic squashed between two icebergs‹ is mirrored in former Canadian Prime Minister Pierre Elliott Trudeau's statement that »[l]iving next to [the United States] is in some ways like sleeping with an elephant. No matter how friendly and even-tempered is the beast, [...] one is affected by every twitch and grunt« (qtd. in Andreas 2005: 462). In these metaphors, Canada always assumes the role of the somewhat helpless junior partner that is unable to appropriately deal with its superior.

McLuhan paints a much more positive picture of the almost ideal opportunities Canadians have thanks to their southern neighbor, as he stresses »the human scale of the small country and [...] the immediate advantages of proximity to massive power« (1977: 228). As becomes evident from McLuhan's statement, nationalism and transnationalism are interwoven. Atwood makes a case for the necessity of transcending national borders and embracing a transnational or even humanist approach:

> But we are all in this together, not just as citizens of our respective nation states but more importantly as inhabitants of this quickly shrinking and increasingly threatened earth. There are boundaries and borders, spiritual as well as physical, and good fences make good neighbours. But there are values beyond national ones. Nobody owns the air; we all breathe it. (1982: 392)

In this way, Atwood clearly appeals to an ecological, border-crossing awareness that is a key transnational and truly global issue.

»I Am Canadian«: Multiculturalism, Nationalism, and Transnationalism

The nation-state of the twenty-first century oscillates between national unity, cultural and ethnic pluralism, and simultaneous processes of diversification. Due to the, as some perceive it, specter of transnationalism, nationalism persists and is even on the rise. Globalization is thus closely associated with increasing regionalization. Consequently, borders still matter in the age of transnationalism, though borders are plural and can take on multiple meanings and functions. In »The Written Line«, Russell Brown distinguishes between five different main functions of the Canada–U.S. border—the border as dividing line, the border as sanctuary line, the border as abyss, the border as interface, and the border as margin (1991: 1–27). These roles are fluid and depend on several factors, such as the border crosser. Canada can be described as a ›safe haven‹ for many different groups of people (Gecelovsky 2007: 521). Historically speaking, draft resisters during the Vietnam War may be mentioned alongside other groups of people crossing into Canada, a place of freedom, by crossing the Canada–U.S. border.

Canada is »a land of multiple borderlines, psychic, social, and geographic« (McLuhan 1977: 244). This notion corroborates diplomat Hugh L. Keenleyside's earlier conviction that »[t]he boundary between Canada and the United States is a typically human creation; it is physically invisible, geographically illogical, militarily indefensible, and emotionally inescapable« (qtd. in Loucky/Alper 2008: 12). To address this emotional inescapability, the border between Canada and the United States as a symbol of Canadian distinctiveness, sovereignty, and national identity plays a vital role. Canadian national identity entails the policy of multiculturalism and the idea of a pluralist society. However, multiculturalism is not seen as an absolute good or a panacea for countering disunity. The »compromise policy of multiculturalism within a bilingual framework« (Ernst/Glaser 2010: 8) has advocates and detractors such as Francophone Quebecers and indigenous peoples. In terms of identity, Till Kinzel contextualizes Canadians' sense of ambiguity and alienation historically: »Canada's special situation as first a colonial and then a postcolonial political entity historically entailed an increasing feeling of alienation that had to do with Canada's seemingly precarious political existence in the shadow of its great imperial neighbour, the United States« (2010: 61).

As a result, the border remains essential for the Canadian psyche, because of self-identification by delineation in light of many similarities with the United States. This phenomenon holds particularly true for Anglophone Canadians, who »[d]espite—or rather *because of*—the essential similarities between the two societies« subscribe to »differentiation« (Bow 2008: 342). Brian Bow contends that this process of differentiation can be found in comparable groupings of two countries, one being smaller and the neighboring country much bigger. The smaller countries want to prove »to themselves and to others, that they are separate and different from a larger, very-similar neighbor, as in New Zealand (vis-à-vis Australia), Austria (vis-à-vis Germany), and most of Central America (vis-à-vis Mexico)« (Bow 2008: 342). Bow alludes to Freud's concept of narcissism of small differences and argues that Canada is prone to such a narcissistic need due to »the relatively shallow and profoundly contested nature of the country's sense of national identity« (2008: 342). Therefore, Canadian attempts in national identity construction seek a point of reference in a ›contrapuntal‹ fashion by which Canadians' own voice is related to another voice, yet nonetheless being distinctly present as in a North American chorus. Underscoring Canadian-American differences might appear anti-American, but Andrew Cohen (2007), for instance, labels his fellow Canadians ›Ameri-skeptics‹ as opposed to full-fledged anti-Americans. Most Canadians are merely critical and afraid of American dominance and consequently feel ambivalent and skeptical toward the United States, but they are by no means hostile. While still using the term ›anti-Americanism‹ with regard to Anglophone Canada, Bow also acknowledges the simultaneous presence of a softening or positive effect thanks to familiarity and »metaphors of family relations« (2008: 342). These kinship metaphors are literally inscribed on the Peace Arch, located in the binational Peace Arch Park in Blaine, Washington, and White Rock, British Columbia: »›Children of a Common Mother‹ and ›Brethren Dwelling Together in Unity‹« (Washington State Parks 2012: par. 1).

Seemingly despite, but more accurately because of, blurred cultural distinctions and close ties between Canada and the United States, patriotism and nationalism prove to enjoy lasting popularity in Canada. One of the most prominent examples of how the appeal to Canadian nationalism is used, even instrumentalized, in the media and popular culture is Joe Canadian's ›Rant‹ (Molson Canadian 2000). This famous Molson Canadian beer advertisement proved a »national sensation« (MacGregor 2007: 64), enjoyed enormous success, and led to many imitations (cf. CBC Digital Archives 2000). Bordering is openly played out in this staging of Canadian patriotism, drawing heavily on preconceived notions and stereotypes of Canada and Canadians in the public imagination. In the commercial, a friendly, white, and Anglophone Joe Canadian enters a stage and explains the political, cultural, and linguistic differences between Canadians and Americans. Starting off rather sheepishly, Joe ends in a powerful crescendo with the tagline: »I am Canadian.« In the commercial, the Canadian flag as well as other Canadian symbols are projected on a screen behind the speaker underlining the contrasts between Canada and the United States. According to Garry Sherbert, the Molson ›I am Canadian‹ commercial shows »›Joe Canadian‹ mockingly defending Canadian cultural distinctiveness against the assimilation and misunderstanding of Americans« (2006: 7). The commercial thus highlights that Canadians' greatest fear is assimilation through American influence and domination in the cultural, economic, and political spheres.

Others see the ad in a less innocent way as going against Canada's very nature—a country of benign and more subtle patriotism. These critics characterize the commercial as »an overtly, even belligerently, patriotic message which struck a chord in a country, that [...] is supposed to be distinguished precisely by its absence of overt patriotism« (Millard/Riegel/Wright 2002: 15). The ad is thus seen as indicative of a societal change and not merely an »aberration« (Millard/Riegel/Wright 2002: 15). Sugars elaborates on the paradoxical nature of the ›Rant‹: »On the one hand, the ad is committed to a defence of a distinct Canadian national-cultural identity; on the other, it accomplishes this message through its evocation of an identity that may not exist« (2006: 123). The ›Rant‹ clearly demonstrates the lingering need of Canadians to delineate themselves from Americans and to reaffirm their own distinctiveness and elusive identity in the absence of strong affirmative identity markers. Once again, self-identification happens by negation of sameness with Americans and American culture. Ironically, Molson has been bought by an American beer company and thus the famous ad could be called »just another American product« (MacGregor 2007: 64).

The very expression of Canadian patriotism has thus become absorbed into the U.S. mainstream. Overall, Roy MacGregor reinterprets the »I am Canadian rant« as an »I am *not* American rant« and highlights the pervasiveness of an identity definition based on negation (2007: 64). Nonetheless, Canadians fall prey to the very things they criticize in American behavior: »Waving the Canadian flag wherever one goes on grounds that it stands for the ›quiet North American‹ is like entering a library and loudly announcing one's silence« (Millard/Riegel/Wright 2002: 21). This contradiction and double standard is symptomatic of constructed differentiation and Othering.

The significance of the ›Rant‹ transcends the immediate reverberations as a cultural phenomenon in the media, since the ad »demonstrates the ways in which, in the late twentieth [and, likely even more so, the early twenty-first] century, Canada is a *contested* space (contested by the apparitional others of US dominance, First Nations claims, Quebec separatism, and ›foreign‹ immigration)« (Bociurkiw 2011: 2). Canada, a decade into the twenty-first century, still needs to negotiate spatial controversies, whether aboriginal land claims or separatist efforts. However, as a sovereign country, Canada enjoys a great international reputation. Consequently, Canadians can self-confidently feel less like ›Romanized Gauls‹, but rather embrace and celebrate the plurality and diversity that is characteristic of Canadian identity.

REFERENCES

Ackleson, Jason (2009): »From ›Thin‹ to ›Thick‹ (and Back Again?): The Politics and Policies of the Contemporary US–Canada Border«, *American Review of Canadian Studies* 39 (4), pp. 336–351.
Andreas, Peter (2005): »The Mexicanization of the US–Canada Border: Asymmetric Interdependence in a Changing Security Context«, *International Journal* 60, pp. 449–462.
Atwood, Margaret (1972/1979): *Surfacing*. London: Virago.
——— (1982): *Second Words*. Toronto: Anansi.
——— (2005/2006): *Curious Pursuits: Occasional Writing*. London: Virago.
Bociurkiw, Marusya (2011): *Feeling Canadian: Television, Nationalism, and Affect*. Waterloo: Wilfrid Laurier University Press.
Bow, Brian (2008): »Anti-Americanism in Canada, Before and After Iraq«, *American Review of Canadian Studies* 38 (3), pp. 341–359.
Brown, Russell (1991): »The Written Line«, in Robert Lecker (ed.), *Borderlands: Essays in Canadian–American Relations*. Toronto: ECW Press, pp. 1–27.
Campbell, Alastair/Cameron, Kirk (2006): »›The North‹: Intersecting Worlds and World Views« in Garry Sherbert/Annie Gérin/Sheila Petty (eds.), *Cana-*

dian Cultural Poesis: Essays on Canadian Culture. Waterloo: Wilfrid Laurier University Press, pp. 143–173.

CBC Digital Archives (2000): »The Rant«, CBC Digital Archives [online], n. d., http://www.cbc.ca/archives/categories/economy-business/business/selling-suds-the-beer-industry-in-canada/i-am-canadian.html. 30 July 2012.

Cohen, Andrew (2007): The Unfinished Canadian: The People We Are. Toronto: McClelland & Stewart.

Druick, Zoë (2006): »Framing the Local: Canadian Film Policy and the Problem of Place« in Garry Sherbert/Annie Gérin/Sheila Petty (eds.), Canadian Cultural Poesis: Essays on Canadian Culture. Waterloo: Wilfrid Laurier University Press, pp. 85–98.

Ernst, Jutta/Glaser, Brigitte (2010): »Introduction«, in Jutta Ernst/Brigitte Glaser (eds.), The Canadian Mosaic in the Age of Transnationalism. Heidelberg: Universitätsverlag Winter, pp. 7–17.

Gecelovsky, Paul (2007): »Northern Enigma: American Images of Canada«, American Review of Canadian Studies 37 (4), pp. 517–535.

Hutcheon, Linda (1988): The Canadian Postmodern: A Study of Contemporary English-Canadian Fiction. Toronto: Oxford University Press.

Kinzel, Till (2010): »Transcending Multiculturalism? Neil Bissoondath and the Question of Canadian Identity«, in Jutta Ernst/Brigitte Glaser (eds.), The Canadian Mosaic in the Age of Transnationalism. Heidelberg: Universitätsverlag Winter, pp. 61–74.

Kroetsch, Robert (1997): »Disunity as Unity: A Canadian Strategy«, in Ajay Heble/Donna Palmateer Pennee/J.R. (Tim) Struthers (eds.), New Contexts of Canadian Criticism. Peterborough: Broadview Press, pp. 355–365.

Laxer, James (2003): The Border: Canada, the U.S. and Dispatches from the 49th Parallel. Toronto: Doubleday Canada.

Loucky, James/Alper, Don (2008): »Pacific Borders, Discordant Borders: Where North America Edges Together«, in James Loucky/Don Alper/J. C. Day (eds.), Transboundary Challenges in the Pacific Border Regions of North America. Calgary: University of Calgary Press, pp. 11–38.

MacGregor, Roy (2007): Canadians: A Portrait of a Country and its People. Toronto: Penguin Books.

McLuhan, Marshall (1977): »Canada: The Borderline Case«, in David Staines (ed.), The Canadian Imagination: Dimensions of a Literary Culture. Cambridge: Harvard University Press, pp. 226–248.

Millard, Gregory/Riegel, Sarah/Wright, John (2002): »Here's Where We Get Canadian: English-Canadian Nationalism and Popular Culture«, American Review of Canadian Studies 32 (1), pp. 11–34.

Molson Canadian (2000): »The Rant«, YouTube [online], 18 May 2006, http://www.youtube.com/watch?v=pXtVrDPhHBg. 30 July 2012.

Morrison, Katherine L. (2003): *Canadians are not Americans: Myths and Literary Traditions*. Toronto: Second Story Press.

New, W. H. (1998): *Borderlands: How We Talk About Canada*. Vancouver: University of British Columbia Press.

Nord, Douglas C. (2011): »Discourse and Dialogue between Americans and Canadians: Who is Talking to Whom?«, *American Review of Canadian Studies* 41 (4), pp. 391–401.

Sherbert, Garry (2006): »Introduction: A Poetics of Canadian Culture«, in Garry Sherbert/Annie Gérin/Sheila Petty (eds.), *Canadian Cultural Poesis: Essays on Canadian Culture*. Waterloo: Wilfrid Laurier University Press, pp. 1–23.

Sugars, Cynthia (2006): »Marketing Ambivalence: Molson Breweries Go Postcolonial«, in Garry Sherbert/Annie Gérin/Sheila Petty (eds.), *Canadian Cultural Poesis: Essays on Canadian Culture*. Waterloo: Wilfrid Laurier University Press, pp. 121–141.

Washington State Parks (2012): »Interpretation and History of Peace Arch«, *Washington State Parks* [online], n. d., http://www.parks.wa.gov/parks/?selected park=Peace%20Arch&subject=interp. 30 July 2012.

Marginalized Cultural Spaces

Spaces of Native American Ghostliness in Thomas Pynchon's *Mason & Dixon*

DIANA BENEA

In the fifteen years since its publication, Thomas Pynchon's novel *Mason & Dixon* has offered fertile ground for various analyses, from studies investigating the novel's narrative strategies and related questions of narrative reliability and their larger epistemological implications (Duyfhuizen 2000) to inquiries into the novel's historical method (Burns 2003), and readings informed by such critical methodologies as postcolonial theory (Smith 2003), queer studies (Sears 2003), and postnationalist theories (Pöhlmann 2010).[1] So far, two edited collections of critical essays exclusively dedicated to *Mason & Dixon* have been published, Brooke Horvath and Irving Malin's *Pynchon and Mason & Dixon* (2000) and Elizabeth Jane Wall Hinds' *The Multiple Worlds of Pynchon's Mason & Dixon: Eighteenth-Century Contexts, Postmodern Observations* (2005), along with a number of book-length studies, numerous articles collected in essay collections on Pynchon's work, and journal articles.[2] Such critical proliferation should not be surprising given the scope of the novel, in the good old fashion of Pynchon's other encyclopedic works, such as *V.* (1963) and *Gravity's Rainbow* (1973), and its exploration of hugely ambitious themes, which are mainly related to the legacy of modernity, such as rationalism, colonialism, nationhood, and the process of nation-making. It is the trope of America that offers an allegory of modernity in the novel, and critical interest may likewise be attributable to the work's treatment of this grand theme: the beginnings, the meaning, the (broken) promise of America. In this sense, quite a few studies have taken a closer look at representations of America in the novel, from investigations of the American West »as subjunctive space« (McHale 2000: 44) to readings of the titular characters as performing an ironic revision of the myth of the American Adam (Greiner 2000).

If *Mason & Dixon* is like the card table in the opening scene of the novel, »with so many hinges, sliding Mortises, hidden catches and secret compartments that neither the Twins nor their Sister can say they have been to the end of it« (Pynchon 1997: 5–6)—an inspired mise-en-abyme for the multi-layered structure of the novel—then one of the ›hinges‹ that has been insufficiently investigated is the figure of the ghost and the cultural and political work it performs in the novel.³ In her influential study *Ghostly Matters: Haunting and the Sociological Imagination*, Avery F. Gordon analyzes the ambivalent figure of the ghost as »the merging of the visible and the invisible, the dead and the living, the past and the present« (1997/2008: 24), as a political figure pointing to the ways in which previous repressions, forced invisibilities, and unresolved historical injustices erupt in the present, making themselves visible in social life, and demanding the attention of those haunted. Drawing on Gordon's exploration of ghosts and the nature of haunting, as well as on other works operating within the framework of what Jeffrey Andrew Weinstock (2004) has called ›the spectral turn‹ of cultural theory, this chapter aims to analyze one particular space of ghostliness and invisibility in Pynchon's text, that of Native Americans, and the novel's general articulation of America as a haunted space troubled by revenants from its history of repression and dispossession.

Mason and Dixon's story of surveying the boundary of Pennsylvania and Maryland between 1763 and 1767, the line that would later bear their own names and serve as the demarcation between North and South and, by extension, as the symbol of a series of binary oppositions, such as freedom/slavery, culture/nature, and innovation/tradition, is a story famous enough not to be recounted here in great detail. Interestingly, the story is narrated from the privileged perspective of a later time, the Christmas Advent of 1786, a post-revolutionary time that sees the new American nation troubled by »wounds bodily and ghostly, great and small, [that] go aching on, not ev'ry one commemorated—nor, too often, even recounted« (Pynchon 1997: 6). The unnamed extra-diegetic narrator of the first few paragraphs of the novel soon gives the floor to one of the characters in his story, Revd Wicks Cherrycoke, a self-declared »untrustworthy Remembrancer« (1997: 8) who will narrate the story of Mason and Dixon, not only recounting the events he witnessed first-hand, but also imagining those he was absent from. Enjoying an extended stay at his sister's home in Philadelphia, Cherrycoke is tolerated as a guest as long as he can entertain the children of the family with various tales, specifically selected for their »moral usefulness« (1997: 7). It is in response to the children's request of »a Tale about America« (1997: 7), preferably an exotic one featuring Indians and Frenchmen, that Cherrycoke begins to spin his long yarn about Mason and Dixon's adventures, at first in Africa to observe the Transit of Venus and later in America to survey the line through the heart of American wilderness. Of particular interest for the present investigation are the ways in which Wicks' narration makes visible those ghostly wounds, unremembered and unrecounted, that trouble post-revolutionary America. Given that the historical events most referenced in the American section of the book are related to the French and Indian War and its aftermath, it becomes clear that one of the wounds alluded to by the narrator lies in America's treatment of its Indigenous population.

Mason & Dixon is populated with an assortment of ghosts and ghostly characters, located on the various narrative levels of the text, diegetic (the ghosts roaming through the LeSpark residence at night) and hypodiegetic (Rebekah, the soldier Dieter in St. Helena, the Cock Lane ghost in London, the ghostly Malays at the Cape, the invisible African slaves at the Lepton iron-plantation, the spectral Indians in the South Mountain region etc.). Commenting on the productivity of this trope, Kyle Smith notes that the ghost functions as a signifier of the »forgotten forces of gender, race, and class« repressed in the process of nation-making and of their »history of peripheralization and elision« (2003: 193). Smith further argues that such over-inclusiveness might be problematic, as it unavoidably implies the loss of specificity of those »various Others« (2003: 194). While this may apply to some extent to the larger social groups represented as ghosts, it is also true that some of the more idiosyncratic ghosts of the novel can serve to illumine the work performed by others.

A brief discussion of the most prominent ghost in the novel, that of Mason's dead wife, Rebekah, is in order here, as it will offer some valuable insights for the analysis of collective ghosts. The ghost of Rebekah visits Mason at various points in the story, as a reminder of his betrayed loyalties to his family for the realm of »pure Mathesis« (Pynchon 1997: 134). Mason is thus condemned to walk the face of the earth as a »Commissioner of Unfinish'd Business« (1997: 171) ever attending to Rebekah's ghost, obsessing over the possibility of becoming reunited with his wife after his death. Tormented by Rebekah's ghost for the most part of the novel, Mason considers it his duty to prevent her passing into forgetfulness, and therefore tells Dixon, at great length and in as great detail as possible, the story of how they first met at an annual feast in their local community, going to considerable lengths so as to »avoid betraying her« (1997: 167). Later on in the novel, Cherrycoke comments upon Mason's intricate tales in a similar manner, describing them as »a way for him to be true to the sorrows of his own history [...], a way of keeping them safe, and never betraying them« (1997: 316). Attending to the ghost implies not only an act of remembrance, but also the ethical injunction of telling the story of the ghost, making it part of the discourse of the living, and thus, in a sense, bringing it back to life. If these are the terms of Mason's relation to an individual ghost, that of Rebekah, one of the central questions of the novel is, I believe, what happens in the case of collective ghosts haunting larger communities. How can they be represented so as to avoid betraying them and how can national communities remain ›true to the sorrows of their own history‹, to paraphrase Mason, precisely by acknowledging the wounds deeply woven into the fabric of the nation?

In her introduction to *The National Uncanny: Indian Ghosts and American Subjects*, Renée L. Bergland traces a shift in ghost belief starting with the Enlightenment, arguing that individual, family ghosts lost some of their hold on people's imagination in this age, while the collective ghosts that haunted (emerging) national communities became increasingly prominent (2000: 9). Bergland thus posits an insightful link between nations as imagined communities (drawing on Benedict Anderson) and hauntedness, suggesting that any process of nation-making implies the ghosting of the communities that do not fit the idea of the nation, that is, the creation of »national ghosts« (2000: 17). The ghost thus becomes a political entity, and haunting a type of political consciousness of the repressions inherent in the construction of the nation. Furthermore, the spectralization of Indigenous people is related to the rise of Enlightenment in other ways as well, as Natives were considered to be »incapable of, or even incompatible with modernity« (Boyd/Thrush 2011: xviii), hence their ghosting in this period. Set in the age of Enlightenment, or the ›Age of Reason‹, as it is frequently referred to in the novel when the protagonists wish to question occurrences that do not conform to their rationalist assumptions, *Mason & Dixon* offers, in its American section, the picture of a nation in the making, of »something styling itself ›America‹, coming into being« (Pynchon 1997: 405), while its national ghosts keep piling up in the background.

It would be interesting to see how the titular characters imaginatively constructed America before actually undertaking the expedition, especially whether they entertained any exceptionalist expectations, and how their American experience either confirmed or subverted their Americas of the mind. Tellingly, before leaving for America in 1763, Mason and Dixon do engage in a debate prompted by their different conceptualizations of America. While the more mystical Mason pictures it as nothing but a British colony surrounded by uncharted territories of »Savages, Wilderness« (1997: 248), where all notions of safety must be suspended, the pragmatic Dixon wonders whether they will be encountering a land that is any different from their previous destination, the Dutch colony of Cape Town, where they became acquainted with the atrocities of a slave society, haunted by a »Collective Ghost of more than household Scale,— the Wrongs committed daily against the Slaves, petty and grave ones alike, going unrecorded, charm'd invisible to history« (1997: 68). Not only does Dixon expect to see the collective ghost of slavery again, but another form of oppression as well, that of Natives, having been told that Americans do not hesitate to kill the inhabitants of the lands where they wish to settle.

Shortly after their arrival in America, the topic of Native Americans is brought up in a casual conversation with Col. Washington at his Mount Vernon residence. Discussing the current progress of settlement, the two surveyors learn from Washington that nothing deters Americans from further expanding west. It is only after Mason and Dixon's inquiry about General Bouquet's Proclamation forbidding settlement west of the Allegheny Ridge Line that Washington mentions the Indians, questioning at the same time the innocence of Bouquet's motives. A reckless Washington even argues that he will not refrain from hiring mercenaries to do the fighting against the Indians if that proves a more efficient way of exterminating them. This fictional representation of Washington is in line with the politics vis-à-vis the Natives of the real George Washington. Commenting on the huge federal expenses on Indian wars and treaties during Washington's presidency, Bergland suggests that Americans »laid claim to citizenship by joining in the public financing of, or active military duty« (2000: 50) in the war against Indians. American identity in the 1790s was therefore constructed in relation to Native Americans, projected as a negative Other that had to be eliminated. What is striking in the approach of Pynchon's Washington is that he does not even bother to construct a discourse that would legitimize the war against this ethnic group, in the tradition of typical colonizers' discourse postulating the savage, backward nature of the colonized as motivation for their subjection, considering instead that he is *a priori* entitled to dispense with the lives of those who might stand in the way. In Washington's account, Indians therefore figure at first as an absence, and then as eradicable entities, systematically threatened into non-existence by a relentless politics of extermination. Certainly, constructing the Indian as either absent or vanishing in the near future is a discursive technique meant to reinforce the project of their actual political removal. Washington here offers an example of what Bergland (2000: 40) argues is the intertwining of two motifs in the public discourse of the early Republic, but the idea is perfectly applicable to the years preceding the Revolutionary War as well: the birth of the American nation and the death of the Native American. The Native American ghost thus becomes a »corollary to American nationalism« (Bergland 2000: 59).

The next mention of Natives in *Mason & Dixon* is in the context of the Paxton Boys' Massacre in Lancaster, when twenty-six Conestoga Indians, survivors of a previous attack by the same vigilante group, are killed. Although failing to register the significance of the event at first, confining themselves to remarks about the irony of whites being the perpetrators of such shocking acts of barbarism, both at the Cape and in America, thus focusing on the aggressors and not on the victims, Mason and Dixon do decide to visit the site of the massacre later in the novel. Appropriately saddened and outraged, Mason comments:

Acts have consequences, Dixon, they must. These Louts believe all's right now,—that they are free to get on with Lives that to them are no doubt important,—with no Glimmer at all of the Debt they have taken on. That is what I smell'd,—Lethe-Water. One of the things the newly born forget, is how terrible its Taste and Smell. In Time, these People are able to forget ev'rything. [...] In America, as I apprehend, Time is the true River that runs 'round Hell. (Pynchon 1997: 346)

Dixon reacts even more violently, praying that justice be restored, that perpetrators be appropriately punished and that he »be spar'd the awkwardness of seeking them out [him]self« (1997: 347). However, Mason and Dixon's justice-seeking discourse is in no way matched by their actions: as soon as they return from the site, they wish to »disengage« (1997: 348) from Lancaster immediately, finding the routine of their scientific enterprises and the comfortable life at the Harlands, their temporary hosts, preferable to experiencing such emotional turmoil. Just as they are leaving Lancaster, they hear »a Voicing disconsolate, of Regret at their Flight« (1997: 348).

The reader later learns that this is not Mason's first disengagement from violent social events occurring in his spatial immediacy: in 1756, when General Wolfe was suppressing the Weaver's rebellion in Mason's home town of Stroud, he chose to leave the place to work as the Royal Astronomer's assistant, his fierce support for the Weaver's cause notwithstanding. In choosing to leave the place at once, Mason and Dixon drink from Lethe water themselves, just as the newly born nation erases the memory of such atrocious events from its public consciousness and marches on triumphantly. The ghostly voice the two surveyors hear as they depart, expressing regret at their failure to translate their discourse into action, is a signifier of anxiety and guilt over their complicity in this national exercise of forgetfulness. Haunting represents the settlement of the debt the whole nation has contracted by allowing such events to happen and to be forgotten. A modern nation living in the now, unencumbered by the burden of history, will later be disturbed by the atemporality of such revenants.

The ghostly apparitions will return as Mason and Dixon start the West line of the project, advancing toward the Warrior Path, an ancient Indian road which represents the boundary they are not supposed to cross. Ghostly figures lurking in the woods, moving in and out of visibility, become a regular presence, thus instilling general confusion and anxiety among Mason and Dixon's crew. The relation between dispossessor and dispossessed, possession understood here in both its physical and supernatural dimensions, is thus more complicated than it would seem at first sight: just as they advance west, taking possession of the Natives' lands with their line, so do the ghosts begin possessing Mason and Dixon's minds. As they gradually approach the sacred Warrior Path, the ghosts of the forest become ever more vocal, at first in an imploring tone, »No ... no more ... no further« (1997: 634), later acquiring threatening connotations, as what they style »the Presence itself« tells them, »You are gone too far, from the Post Mark'd West« (1997: 635).

The trope of the ghost is highly ambivalent in this context, signifying both the powerlessness of Natives to stop the progress of the surveyors through their lands, and their firm emplacement as figures of fear in the imagination of those who are displacing them. Mason, for instance, attributes his anxiety toward the Natives to their elusiveness and constant slippage in and out of sight, thus violating the rational categories of visibility and invisibility, and to their special connection to the land they inhabit, that quality of belonging »so *without separation, to this Country cryptick and perilous*« (1997: 648), as if they were an extension of their land.

The most substantial representation of Natives occurs, however, in the last stages of marking the West line, when representatives of various Indian tribes accompany the surveyors, so as to show them the precise location of that sacred meeting-ground, the Warrior Path, that is, the terminus of their enterprise. The Natives and the surveyors engage in some comparative ethnography of their respective worldviews, discussing such topics as the location of their gods, the names of stars and the possibility of life in other worlds, pointing out differences as well as similarities that undermine the surveyors' expectations of absolute Otherness. During these exchanges with Indian tribe members, the issue of supernatural haunting is articulated for the first time from a Native's point of view:

Long before any of you came here, we dream'd of you.[...] You had Powers and we respected them. Yet you never dream'd of us, and when at last you saw us, wish'd only to destroy us. Then the killing started,— some of you, some of us,—but not nearly as many as we'd been expecting [...].Instead, you sold us your Powers,—your Rifles,—as if encouraging us to shoot at you,—and so we did, tho' not hitting as many of you, as *you* were expecting. Now you begin to believe that we have come from elsewhere, possessing Powers you do not [...]. Those of us who knew how, have fled into Refuge in your Dreams, at last. Tho' we now pursue real lives no different at their Hearts from yours, we are also your Dreams. (Pynchon 1997: 663)

The reader is thus presented with the Native interpretation of the white-Native divide and with an explanation of the Natives' haunting of the whites. Not generous enough to imagine the existence of other nations on the lands they wished to inhabit, settlers will nevertheless have to later accommodate the Natives who ›settle‹ in their imagination. In an interesting reversal of power relations, the colonizers' dreams are thus colonized by those whom they dispossessed and corrupted (whites introducing Indians to advanced weapons will become a recurrent motif in the novel). Furthermore, the settlers fail to see haunting as a result of their violence toward the Natives, assigning such phenomena to the Natives' supernatural powers, to their ›coming from elsewhere‹, that is, to their Otherness. What the whites forget is that ghosts are never immaterial presences from another world, but very much related to the palpable world they disturb. As Gordon argues, »to be haunted is to be tied to historical and social effects« (1997/2008: 190).

Representations of Native Americans in this section are by no means idealized. There are various mentions of tribes engaging in violent warfare with one another, as well as of cruel killings of white settlers. Through his balanced portrayal of the Natives, represented both in their Otherness and in their commonalities in relation to the whites, Pynchon manages to avoid the trap of producing an Indian image that non-Native audiences might fetishize for its benign exoticism or noble savagery or comfortably appropriate as a slightly different version of their own.

Reaching the Warrior Path, the crew discontinues the West line, although both surveyors articulate, at various points, their strong desire to go on with the project, beyond the Indian road, either by negotiating with the Indians or bribing them with spirits, as Dixon jokingly suggests. In fact, the two wish to proceed with the line not so much out of disregard for the Natives, but as a result of having become engrossed in visions of the Eternal West, or »Rapture de West« (Pynchon 1997: 670). Constructed as a mythical space of fantastic creatures and golden cities, where personal freedom and the power of imagination reign supreme, Mason and Dixon's unbounded and unmapped West can be interpreted as a heterotopia far removed from modern America and its piles of ghosts. Significantly, the continuation of the line does occur in the subjunctive realm, as the narrator imagines, in the last chapter of the American section, their further adventures beyond the Ohio River, where they have to confront the real-life consequences of their dividing line. If Indians escape the violation of the interdiction to cross the Warrior Path in the space of the narrated, this violation is performed nevertheless in the realm of the disnarrated, the hypothetical, and the counterfactual. The crossing of the Warrior Path amounts to another instance of brushing aside the will of the Indians, that is, of ghosting them, whose agent in this case is the extra-diegetic narrator.

Indians emerge for the last time in the very final paragraph of the novel. Although Dixon was the one ever fascinated with America, it is Mason who returns to Philadelphia and settles there after many years spent at home, in England, following their surveying of the boundary. After Mason's death, his wife and younger children decide to return to England, while his eldest sons, William and Doctor Isaac, Rebekah's sons, choose to stay in America and »be Americans« (1997: 772). In the final paragraph of the novel, the sons reminisce (or perhaps dream?) about their desire as children to accompany their father on his American expedition and, in the absence of such direct experience of America, recall the ways in which they imagined that mysterious space. What is significant here is how the past of the narration swiftly modulates into the present:

»Since I was ten,« said Doc, »I wanted you to take me and Willy to America. I kept hoping, ev'ry Birthday, this would be the year. I knew next time you'd take us.«
»We can get jobs,« said William, »save enough to go out where you were,—«
»Marry and go out where you were,« said Doc.
»The Stars are so close you won't need a Telescope.«
»The Fish jump into your Arms. The Indians know Magick.«

»We'll go there. We'll live there.«
»We'll fish there. And you too.« (Pynchon 1997: 773)

Such an ending might seem somewhat incongruent in the general context of the book, as it works to reaffirm the myth of America, gesturing toward the perpetual need to construct it in utopian terms, as a space of desire. Some critics have, indeed, taken Pynchon to task for the »forgiving pastoralism« (Schaub 2000: 200) of the endings of both *Vineland* (1990) and *Mason & Dixon*, which lack the sense of urgency dominating his previous works, working instead in the genre of »nostalgic (or bourgeois) tragedy« (Schaub 2000: 201). The shift in tonality also seems to be illustrated by the final representation of Native Americans in the text, namely by the transition from Indian ghosts as figures of terror and lament to the harmless, magic-practicing Indians. My suggestion is to read this final image of the Natives in conjunction with an earlier exchange between Mason and Dixon, from their days at the Cape. Dixon's answer to Mason's inquiry about the various sorts of magic practiced by the Malays reads: »It may content us, as unhappy grown Englishmen, to think that somewhere in the World, Innocence may yet abide,—yet 'tis not among these people. All is struggle,—and all but occasionally in vain« (Pynchon 1997: 67). Implicit in the final passage of the novel is a critique of that self-indulgent fascination with Otherness (and the generous mourning at the thought of its vanishing) that actually conceals the failure to address the unresolved ›struggles‹ and the social violence committed against these Others. The image of Indians as noble savages practicing magic might be comforting, but it obscures such events as the Paxton Boys' Massacre or the Siege of Fort Pitt, that is, a history of violent oppression that the text otherwise makes visible.

In Gordon's view, ghosts should serve not only as reminders of a history of exclusions and disappearances, but also as signifiers of a loss, of »a path not taken« (1997/2008: 64). To be hospitable to the ghost, Gordon suggests, is »to allow the ghosts to let you imagine what was lost that never even existed« (1997/2008: 57), an imaginative effort meant to inspire a future change, an alternative course of history. This trope is reminiscent of a passage in Pynchon's *Gravity's Rainbow* detailing the history of Tyrone Slothrop's ancestor, William Slothrop, who settled in America in the 1630s. In his heretical tract entitled *On Preterition*, the elder Slothrop discusses the category of the Preterite, »the many God passes over,« arguing for the »holiness [of] these ›second sheep‹ without whom there'd be no elect« (Pynchon 1973: 555). More than three centuries later, Tyrone Slothrop wonders whether his ancestor's heresy might have been »the fork in the road America never took, the singular point she jumped the wrong way from« (1973: 555). While the discourse of Native American spectralization in *Mason & Dixon* illuminates significant aspects of the rhetoric of colonialism and the process of nation-making, its value also lies in inviting reflection on the different paths that could have been taken and could still be taken in America's treatment of its Natives and in its general treatment of Others. The Native American ghosts in the novel serve as an index of retrospective national guilt vis-à-vis this ethnic group, as residues of the violence they have experienced, but they should also encourage some prospective thinking, some serious ethical consideration of what can be done differently in the future. Perhaps the ›moral usefulness‹ of Cherrycoke's tale about America resides in the injunction to think about the ghosts around us, to make them visible to our consciousness, and to see them as opportunities for enhanced social justice.

Notes

1 | Work on this chapter was made possible thanks to the financial support of the Sectoral Operational Programme for Human Resources Development 2007-2013, co-financed by the European Social Fund, under project number POSDRU/107/1.5/S/80765.
2 | Interest in the novel, however, has not been confined to the institutionalized spaces of academia and professional literary criticism. One need only browse through the constantly updated *Mason & Dixon* Wiki Page (part of the broader *Pynchon Wiki* created and curated by Tim Ware, which features page-by-page annotations of all of Pynchon's novels), a comprehensive companion for the many references to be found in the novel, for an illustration of the book's continuing appeal to a non-professional readership as well.
3 | Daniel Punday's article »Pynchon's Ghosts« provides the most substantial treatment of the trope, offering an analysis of the central ghost of the novel, that of Rebekah, through the lens of Derrida's *Specters of Marx*. The trope of the ghost is also briefly considered in the last section of Kyle Smith's article »›Serving Interests Invisible‹: *Mason & Dixon*, British Spy Fiction and the Specters of Imperialism« as part of a larger argument about the ideology of whiteness and the strategies employed to oppose it in both *Gravity's Rainbow* and *Mason & Dixon*.

References

Bergland, Renée L. (2000): *The National Uncanny: Indian Ghosts and American Subjects*. Hanover: University Press of New England.
Boyd, Colleen E./Thrush, Coll (2011): »Introduction: Bringing Ghosts to Ground«, in Colleen E. Boyd/Coll Thrush (eds.), *Phantom Past, Indigenous Presence: Native Ghosts in North American Culture & History*. Lincoln: University of Nebraska Press, pp. vii–xl.

Burns, Christy L. (2003): »Postmodern Historiography: Politics and the Parallactic Method in Thomas Pynchon's *Mason & Dixon*«, *Postmodern Culture* 14 (1), online.

Duyfhuizen, Bernard (2000): »Reading at the ›Crease of Credulity‹«, in Brooke Horvath/Irving Malin (eds.), *Pynchon and Mason & Dixon*. Newark: University of Delaware Press, pp. 132–142.

Gordon, Avery F. (1997/2008): *Ghostly Matters: Haunting and the Sociological Imagination*, New Edition. Minneapolis: University of Minnesota Press.

Greiner, Donald J. (2000): »Thomas Pynchon and the Fault Lines of America«, in Brooke Horvath/Irving Malin (eds.), *Pynchon and Mason & Dixon*. Newark: University of Delaware Press, pp. 73–83.

McHale, Brian (2000): »*Mason & Dixon* in the Zone, or, A Brief Poetics of Pynchon Space«, in Brooke Horvath/Irving Malin (eds.), *Pynchon and Mason & Dixon*. Newark: University of Delaware Press, pp. 43–62.

Pöhlmann, Sascha (2010): *Pynchon's Postnational Imagination*. Heidelberg: Universitätsverlag Winter.

Punday, Daniel (2003): »Pynchon's Ghosts«, *Contemporary Literature* 44 (2), pp. 250–274.

Pynchon, Thomas (1973): *Gravity's Rainbow*. New York: Penguin Books.

––––––– (1997): *Mason & Dixon*. London: Vintage.

Schaub, Thomas H. (2000): »Plot, Ideology, and Compassion in *Mason & Dixon*«, in Brooke Horvath/Irving Malin (eds.), *Pynchon and Mason & Dixon*. Newark: University of Delaware Press, pp. 189–202.

Sears, Julie Christine (2003): »Black and White Rainbows and Blurry Lines: Sexual Deviance/Diversity in *Gravity's Rainbow* and *Mason & Dixon*«, in Niran Abbas (ed.), *Thomas Pynchon: Reading from the Margins*. Madison: Fairleigh Dickinson University Press, pp. 108–121.

Smith, Kyle (2003): »›Serving Interests invisible‹: *Mason & Dixon*, British Spy Fiction, and the Specters of Imperialism«, in Niran Abbas (ed.), *Thomas Pynchon: Reading from the Margins*. Madison: Fairleigh Dickinson University Press, pp. 183–198.

Wall Hinds, Elizabeth Jane, ed. (2005): *The Multiple Worlds of Pynchon's Mason & Dixon: Eighteenth-Century Contexts, Postmodern Observations*. Rochester: Camden House.

Weinstock, Jeffrey Andrew (2004): »Introduction: The Spectral Turn«, in Jeffrey Andrew Weinstock (ed.), *Spectral America: Phantoms and the National Imagination*. Madison: University of Wisconsin Press, pp. 3—17.

Getting a Name
Searching for a Mixed-Blood Identity in Sherman Alexie's *Flight*

Madalina Prodan

Ethnicity has been constantly shaping both the physical and the cultural American space, contributing greatly to its specificity.[1] In this context, Indianness plays an important part, becoming a marker for defining Americanness, while, simultaneously and paradoxically, denying Native Americans' right to self-representation, since the dominant discourse has programmatically misrepresented the Indian and muted native voices for centuries. However, during the last few decades, native voices regained control over their representation, repositioning Indianness within the mainstream discourse. One of the most influential Native American writers who contributed greatly to reshaping the American cultural space is Sherman Alexie.

This chapter focuses on Sherman Alexies' *Flight* (2007), a short, but intense novel about different ways of understanding the role of violence in the evolution of its main character, Zits. Alexie's major concern in *Flight* is to reflect on the banality of violence in everyday life and, most of all, on the pervasive complicity of all humans in its existence and perpetuation, which almost break down the mixed-blood orphan. Nevertheless, the author gives his character a chance to begin the process of healing: by witnessing extremely violent historical events, he can face his own personal traumas. Zits' healing can be read as a symbolic social process of healing in white-Indian relationships. Drawing on these observations, the aims of this essay will be threefold: first, I will analyze the character's initial self-positioning in his social milieu, for through his doubly marginalized role as a mixed-blood and an orphan Zits occupies an ambiguous position which allows him to subvert social conventions. Second, I will turn to the character's shift of attitude as he powerlessly witnesses acts of extreme violence while undergoing an unwilling journey through the history of native-white encounters that will eventually help him come to terms with his own situation and work toward reconciliation and social integration. Finally, I would like to suggest the presence of a subtle paradigmatic shift in Sherman Alexie's latest fiction. The Spokane writer's previous affirmation of Indianness endangered by white values shifts to Indian (re)affirmation within what Vijay Prashad has referred to as a ›polycultural‹ environment that allows both Indians and whites to form genuine connections »across perceived lines of racial difference« (Prashad 2001: xii). In other words, while the affirmation of Indianness is still pivotal in Alexie's writing, it does no longer follow the path of Indian *versus* white, but it is rather perceived as Indian *and* white.

Flight centers on Zits, a mixed-blood orphan who has neither ties with his dead mother's white family nor with his Indian father who deserted him when he was born. Having been subjected to the violent and abusive treatment inherent in the foster care system (at least, according to the protagonist) scarred him and apparently doomed him on a downward path to violent (self)destructive behavior: The white character Justice manipulates Zits into shooting people in a bank as a paradoxical solution to his frustration. In the bank, he is shot in the head but, instead of dying, he is transported back in time and forced to witness violent historical events through some fantastical temporary body-shifting process. As he shifts from one body to the next (be it Indian or white), he increasingly controls the body his mind inhabits; from a mere spectator he gradually develops into an active agent.

The novel's protagonist describes himself in a manner characteristic of Sherman Alexie's writing, in which humor and desperation blend together to form a unique mixture: »I'm a fifteen-year-old foster kid with a history of fire-setting, time traveling, body shifting, and mass-murder contemplation. I think I'm a lot more than just dangerous. I think I might be unlovable« (2007: 173). A mixed-blood orphan, estranged from both his Irish and his Indian heritage, moving from one foster home to another, a somehow perfect embodiment of a problem teenager struggling to make sense of the world around him and to find his own place, he tells a story infused with pain, desperation and violence, a story that somehow the writer masters to a reconciliatory happy end.

Flight's opening line—»Call me Zits«—provides an obvious intertextual reference to Herman Melville's *Moby-Dick* (1851). As Alexie stated in an interview, his intention was to »point out that for teenagers the state of their complexion is at least as important as Moby Dick is to Ahab« before explaining that the idea behind naming the novel's central character ›Zits‹ was »to show how much his scars affected him and how much they changed his outlook on the world« (qtd. in Weich 2007: par. 11–12). Indeed, Zits' name symbolizes his inner scarring, his traumatized childhood as an orphan who lost his mother and has never met his father, was physically abused by his aunt's boyfriend, and entered the public system of foster care, moving from one foster home to another since the age of eight. However, there is more to the name than the mere highlighting of Zits' scars, as his name is also a marker of his invisibility: Zits is just an ordinary teenager, unremarkable for society at large, a mere name in the foster care system. Thirdly, by choosing this name, the character consciously positions himself outside the social norm that requires everyone to use a more or less socially accepted and acceptable denomination. And lastly, by calling himself ›Zits‹, the character is negatively inscribing his identity and, at the same time, is dissociating himself (once again) from a society in which physical beauty has normative value.

Sherman Alexie has become one of the most successful ethnic writers, creating »narratives of self-representation that critically question and often radically revise and subvert the dominant culture's conquest narratives and the mass-produced misrepresentations of Native Americans« (Cox 1997: 53–54). In *Flight*, Alexie critiques yet again the destructive potential mainstream representations of Indians still have on Native Americans. Zits has no ties to his native heritage. He has Indian blood from his father, who abandoned him when he was born, but he knows nothing about ›the Indian ways‹, since there is no one to educate him in this respect. Thus, he learns about Indianness from watching TV. Much like John Smith in Alexie's novel *Indian Killer* (1996), Zits acquires knowledge on what it means to be an Indian via the dominant narratives, since these are the only available sources of information he has access to. Yet whereas the conflictive prescription of Indianness in the dominant discourse drove *Indian Killer*'s John Smith to suicide, Zits has the chance to experience first-hand what it means to be an Indian through his travels through time and space. Moreover, through John's desperate attempts to conform to the white normative system, Alexie raises the question of ethnic identity based solely on phenotypic traits. Both Zits and John ›look Indian‹, but they do not know how to ›be Indian‹. In other words, society confines them to certain social personas that lead them on the path of frustration, anger, and eventual self-destruction, since they cannot harmonize the images they receive from the media with daily life. Through these two characters, Sherman Alexie criticizes the pervasiveness of the Vanishing Indian ideology in the media discourse and its traumatic effects on contemporary Native Americans.

The Indians Zits finds in the media discourse are stuck in an indefinite past. In one of his time travels, Zits says, »I'm not exactly sure what year it is. It's tough to tell the difference between seventeenth- and eighteenth- and nineteenth-century Indians« (Alexie 2007: 60). There is a discrepancy between what Indians are supposed to be like according to the dominant discourse, which dooms them in a primitive past, and the present-day natives, who are ordinary Americans, doing the same things like any other American. This tension between imaginary and ›real‹ Indian life is exemplified in a situation in which Zits nostalgically remembers his mother singing to him a Blood, Sweat & Tears song and not some ›traditional‹ Indian lullaby. In *Flight*, Sherman Alexie achieves one of the funniest and most effective deconstructions of the savage, stuck in the past Indian in the Hollywood film industry. In one of his time travels, when Zits inhabits the body of a young Indian, his apparently naive remark exposes the shallowness and falsity of the mainstream representations of Indianness: »Then I solve a mystery: I look under my loincloth. Okay. I know for sure now that Indians didn't have underwear beneath their loincloths« (2007: 63).

The need for some sense of rootedness and belonging determines Zits to acquire more information about his father's ethnic background, for »it makes [him] feel more like a real Indian. Maybe [he] can't live like an Indian, but [he] can learn how real Indians *used to live* and how they're *supposed to live* now« (2007: 12; my emphasis). The artificiality and falsity of the constructed concept of the Indian is made transparent here. Then, later on, the writer explores the influence of the white discourse on Zits, a discourse that renders Native Americans extinct, in the past: »These are how Indians used to be, how Indians are supposed to be. Justice always talked with admiration about Indians like this« (2007: 60).

An interesting aspect of constructing Zits' immersions into the ›authentic‹ Indian world is the fact that Alexie constructs the boy's ›flights‹ based on his limited and distorted knowledge about the Indian ways, which is basically a white perspective, »a white man's artefact,« as Frantz Fanon has it (1952/2008: 6).[2] As Zits confesses at the beginning of the novel: »Everything I know about Indians [...] I've learned from television. I know about famous chiefs, broken treaties, the political activism of the 1960s and 1970s, and the Indian wars of the nineteenth century« (Alexie 2007: 12). Consequently, his flights take him to sites where such events took place. The first mind he temporarily inhabits is that of a white FBI agent who shoots an Indian in the ethnic tensions of the 1960s. Zits literally witnesses the events from the white perspective. Then he jumps into the mind of Indian Boy, who lives at the Indian camp at Little Bighorn in June 1876. Once again, the source of his imaginary incursion in history is the official perspective on the event, since he confesses: »I watched a TV show about it on the History Channel« (2007: 71). Zits' exhilaration when he is transported back in time to this »stinky Heaven« (2007: 65), as he labels the Indian camp, comes from the feeling of finally having a family, even if his discourse has a voyeuristic touch of contemporary white tourism. After all, Zits' perspective in describing the Indian camp is that of a twenty-first century American experiencing the living conditions in the late nineteenth century. Even if his standpoint is initially informed by a white perspective, he quickly surpasses the shallowness of white discourse, sees past the ›inferiority‹ of those living in the Indian camp and empathizes with them.

The next flight is inside the mind of a white veteran tracker, »the best Indian tracker in the entire U.S. Army history« (2007: 84), a man who spent his whole life hunting down Indians and who now has an epiphany of what humanity entails and decides to ›go Indian‹ by helping Small Saint save Bow Boy. He unwillingly recalls the atrocities the Indians committed by desecrating the dead bodies of the colonists. By introducing this episode, Alexie points out that not only white people are to blame for violence and revenge and that, even if they had the excuse of defending their land, Indians and whites alike should explore alternative solutions.

Through the white tracker who ›goes Indian‹, Alexie brilliantly subverts and reinscribes the meaning this expression holds in the American cultural space. As Shari Huhndorf asserts in her study *Going Native: Indians in the American Cultural Imagination*, the temporary adoption of an Indian way of living (›going Indian‹) became »a means of escaping a degenerate and corrupt white world« (2001: 35) in the mainstream narratives of twentieth-century industrialized America. Through this process of going native, the white hero »uncovers his own ›true‹ identity and redeems the European-American society« (Huhndorf 2001: 35). The white hero who goes Indian, temporarily a social outcast, appropriates various aspects of Indianness that would help him define himself and grow, but he eventually surpasses this stage in his life and returns to white society.[3] However, Alexie's white character ›going Indian‹ disrupts this pattern. He is not an Indian lover (on the contrary), and he does not benefit from his act (on the contrary, he endangers his life). The tracker is not so much ›going Indian‹ as ›going humane‹, for his sole reason for helping Small Saint save Bow Boy is provided by an epiphany that allows him to realize that they all share the same humanity.

Afterwards, Zits inhabits Jimmy's mind, a pilot who unknowingly trains a Muslim would-be terrorist (whom he also considers his best friend) how to fly. Adding this story to the architecture of the novel provides depth and universality to temporally and geographically specific ethnic conflicts and also allows the reader to respond even more by evoking the contemporary situation of 9/11 that shocked the United States.

If the first two flights are informed by the projections of the white imaginary (before he has the opportunity to meet real Indians),[4] Zits' perspective undergoes a gradual and subtle dissociation from dominant discourse misrepresentations of the natives. As he keeps shifting from one mind to another, he becomes stronger, his journey acquires a more personal note, and there is less stereotypical representation of Native Americans. There is little wonder that the last mind he inhabits before coming to his own body before shooting the people in the bank is the mind of his alcoholic, beaten-down father. Having gained a better understanding of the larger picture, Zits is finally able to cope with his own personal history, to find answers and to formulate a self-satisfying sense of belonging. Zits' thirst for ethnic knowledge originates in the lack he feels in his personal history. The reason for Zits' immersion into the sea of nativity is to find some sort of connection to his absent father on the one hand and to get some sense of rootedness on the other. The process of healing starts once he has the opportunity to meet his father and to understand the reasons for being abandoned. At the same time, Zits begins to understand what it means to be an Indian in the contemporary world, which is actually the ›real Indian‹, as opposed to the previous image disseminated through the documentaries Zits had seen.

From a postcolonial studies perspective, the colonizer assumes the role of the parent and the colonized is forced into the position of the child. William Cohen traces this process to the countries under British and French domination in the late nineteenth and early twentieth century: »The European imperial powers of the nineteenth and twentieth centuries viewed their African and Asian subjects as men not fully grown, whose destiny had to be guided by the presumably more advanced states of Europe« (1970: 427). The paternalistic attitude of white colonizers toward the colonized Indian tribes was institutionalized in the United States even earlier. For instance, President Andrew Jackson, who signed *The Indian Removal Act* in 1830, described Native Americans as »children in need of guidance« (qtd. in Prucha 1969: 530). Colonization is therefore threefold in Zits' case: he is a child, he is an orphan, and he is a half-breed.

Homi K. Bhabha's concept of ›hybridity‹ proves to be a useful tool in analyzing this ambiguous position Zits finds himself in. In his examinations of postcolonial cultures, Bhabha argues that the identity of the colonized is split: it still bears traces of the former, genuine identity, but it has also engulfed features belonging to the colonizer's culture. The critic labeled this main characteristic of postcolonial cultures ›hybridity‹, a kind of ›in-betweenness‹, the sum of the two cultures, colonizing and colonized. However, the culture of the colonized subverts and revises the dominant culture through mimicry, that is, imitation of the colonizer's culture by employing irony and pastiche. As a collection of cultural discourses, Bhabha regards the colonizer's book as »an unintended vehicle of hybridity and ambivalence« (1994: 103). In its original context, it is a direct product of its culture, but in the colonial context its initial meaning starts slipping away, as it undergoes »an Entstellung, a process of displacement, distortion, dislocation, repetition« (Bhabha 1994: 105). Through this process, the book, used by the colonizer and colonized alike, eventually becomes an ambivalent object, as ambivalent and uncertain as the colonial rule itself. This ambivalence brings about a subversion and revision of the dominant culture, turning the colonizers' weapons against themselves:

> If the effect of colonial power is seen to be the production of hybridization rather than the noisy command of colonialist authority or the silent repression of native traditions, then an important change of perspective occurs. The ambivalence at the source of traditional discourses on authority enables a form of subversion, founded on the undecidability that turns the discursive conditions of dominance into the grounds of intervention. (Bhabha 1994: 112)

Zits occupies a similarly ambivalent position: as an orphan, he is situated both within and outside society. He is quite powerless in making any decisions concerning his fate and the repressive social mechanisms don't allow him much freedom. But, at the same time, because he is positioned outside the family unit as the cornerstone of society, Zits occupies a particularly powerful position due to the potentially disruptive quality of his status as an orphan. As Motoko Sugano points out, »[t]he very illegitimacy, and therefore the powerlessness, of the orphan figure can be read as a power to destabilize the legitimacy of family« (2005: 2). And Zits constantly challenges the power system that places him in the hands of foster parents, running from new foster families time after time and showing great unwillingness to accept any foster parents. At the beginning of the book, when he is ready to flee from the twenty-first foster home, he bitterly confesses: »I'm never in one place long enough to care« (Alexie 2007: 8).

As a problem-causing teenager (an orphan fleeing from foster homes, a shoplifter, an arsonist, and a potential mass-murderer), Zits' marginal position is potentially disruptive. His in-betweenness is further highlighted by his ambiguous status as a mixed-blood. As Zits stresses early on, »I'm not an official Indian. My Indian daddy gave me his looks, but he was never legally established as my father« (2007: 9). On the other hand, he is not quite white, either, since the American psyche adopted Anglo-Saxon mentalities that marked the Irish as ethnically inferior long before the New World was discovered in order to justify their conquest, domination and genocide they subjected the inhabitants of Hibernia to. The Irish had been discriminated against by the English for centuries. For instance, Charles Kingsley, a nineteenth-century English clergyman wrote in an 1860 letter to his wife that the poor peasants affected by the great Irish potato famine were »white chimpanzees« (qtd. in Curtis 1968: 84), which was, in his opinion, worse than being black. The Irish were regarded as an ›inferior race‹ by the Anglo-Saxons, a racialist position that was successfully relocated to the New World. The strongest anti-Irish attitudes (employment and living segregation included) were manifested in the United States after the massive immigration caused by the potato famine mentioned above. Therefore, Alexie's choice for Zits' mother's white European ethnic origins is not entirely arbitrary, for Zits' position within the white racial normative system has a highly disruptive potential. At one point, Zits notes: »I am Irish and Indian, which would be the coolest blend in the world if my parents were around to teach me how to be Irish and Indian. But they're not here and haven't been for years, so I'm not really Irish or Indian. I'm a blank sky, a human solar eclipse« (Alexie 2007: 5). Zits thus has no name, no history, no cultural memory, no sense of belonging. When talking about the colonized status of Native Americans in an interview, Sherman Alexie pointed out that »colonization destroy[ed] family units« (qtd. in Weich 2007: par. 26), which ultimately implies losing connection to history.

»I've been partially raised by too many people« (Alexie 2007: 6), Zits tellingly confesses to the reader.

The lack of connection to history is best conveyed through the motif of the absent father, which is not new in Alexie's fiction, but it gains prevalence in *Flight*, since Zits defines his identity through his father's absence. Alexie stated that the missing parent was »a constant theme of any colonial literature« before explaining the intricacies of displacement as »[t]he killing of your birth father and the substitution of an adopted father. Think of your birth parent being your original culture and your adopted parent being the colonizing culture. In a sense, Native Americans, anybody who's been colonized, they're in the position of an orphan« (qtd. in Weich 2007: par. 21; par. 23).

Alexie's remarks bring to mind Edward Said's theory of filiation and affiliation.[5] According to Said (1983), the orphan enters affiliative relationships in order to fill the lack of a filial relationship. In Zits' case, besides the absence of natural bonds, there are hardly any affiliational relationships to replace them. The sense of betrayal and violence is so much rooted in his life and his traumas scarred him so deeply that he simply does not trust anyone to fill the void. As he confesses: »I learned how to stop crying. I learned how to hide inside of myself. I learned how to be somebody else. I learned how to be cold and numb« (Alexie 2007: 161).

Another reason for not being able to fill this void is his profound distrust in his fellow human beings. The only one who gets close to him is Justice, a white teenager he meets in jail. There is something particularly interesting about Justice, for he can easily be regarded as Zits' alter ego. With translucent skin, as opposed to Zits' scarred complexion, the white teenager seems to have the answers to all the questions and thus represents a rationalization of Zits' anger and frustration. »Hate can be empowering« (2007: 22), Justice tells Zits, a phrase which Alexie's readers are already familiar with, though to different ends.

Justice manages to get close to Zits, to brainwash and to manipulate him into mass-murder on the grounds of bringing his father back through ghost-dancing. The white teenager senses the mixed-blood's feelings of anger and frustration and channels these feelings into the most common human reaction: revenge. In Zits' particular case, revenge takes the form of shooting innocent people in a bank. Justice manipulates Zits into killing on the grounds of the perpetuation of racism against Indians in white society. But the location for doing so is important, since the bank represents the very core of white capitalism and most of the poorest people in America belong to various ethnic minorities. Indians are thus doubly oppressed.

Nevertheless, *Flight*'s main objective seems to be emphasizing the complicity of all human beings in the existence and perpetuation of violence. The initiation Zits undergoes in his flights results in a new awareness of the destiny of mankind: »Maybe we're all lonely. Maybe some of them also hurtle through time and see war, war, war. Maybe we're all in this together« (2007: 158). No one is free of guilt or innocent: »Yes, I'm supposed to kill for Justice. I did it before: a long time ago, a little while ago, a second ago. I don't understand how time works anymore« (2007: 157). As a means of human interaction, violent acts thus transgress ethnicity and have been repeated countless times throughout the history of mankind.

But Zits gradually changes while witnessing extremely violent white-Indian encounters. The memorable events in native–white relations Zits relives are the Battle of Little Bighorn, the political American Indian movement in the 1970s, Indian attacks on white settlements and white incursions onto Indian settlements during the Indian wars, all of which provide the basis for the (mis-)representation of the Indian in dominant discourse. As becomes evident from Zits' journeys through time and space, the driving force behind not only native-white relations, but human relations at large, is hate and the lust for revenge. The result is always death, desecration of bodies, and once again the desire for revenge, a never-ending cycle that leaves no hope for breaking down the legacy of violence. But, as Zits powerlessly witnesses scenes of nauseating violence, he encounters both whites and Indians who have the desire to break up the vicious circle.

The first one to break the revenge-killing circle is a mute Indian boy, whose pacifist intentions in a world filled with violence are symbolically silenced. Zits comments, »He's a child and I'm a child and I'm supposed to slash his throat« (2007: 78). Then, there is Small Saint, a sixteen-year-old white soldier saving Bow Boy, a five-year-old Indian, with the help of a veteran in the middle of white attacks on Indian villages. Together they »outrun the monster revenge« (2007: 97).

The last time travel is inside the mind of a bitter and homeless alcoholic Indian who blames whites for all the wrongs done to Indians. Full of rage and self-hatred spurred by a traumatic childhood, this crushed man turns out to be Zits' father. Most of Zits' frustration and anger can be traced to his inability to deal with being abandoned by his father. Now that he has the opportunity to finally unveil the reasons for his father's behavior, Zits forces his way into his father's memories, only to discover that his father was abused and traumatized by his own father, that running away was the only action he could take in regard to parenthood. Repeatedly forced by his father to say out loud »I ain't worth shit« (2007: 155), Zits' father grows up with feelings of inferiority and impotence, violence and self-hatred, which overpower the basic human desires for love, kinship, and acceptance. Zits' remark »I am my father« (2007: 150) can be interpreted not only as a statement of inhabiting his father's mind but also as providing testament to a newly gained awareness that by renouncing love and following the path of revenge, Zits behaves exactly like his father, probably wasting his life in bitterness, as his father did, hurting the people around him.

He is destined to have his father's fate if he doesn't deal with his traumas, if he allows his anger to control him, and if he stays on the path of violence. Like Billy Pilgrim in *Slaughterhouse-Five* (1972), Zits acknowledges the absurdity of violence, but if Billy Pilgrim is defeated and escapes into alternative realities, Zits manages to pull himself out of the paralyzing pain and change something in himself and, implicitly, in the world around him.

The first step in the process of healing acknowledges the sickness and the need for help. The teenager confesses: »I am tired of hurting people. I am tired of being hurt. I need help« (2007: 161–162). Thus, Zits transforms into Michael. He is no longer invisible. For the first time he is willing to integrate into a world that had shown him only violence. He is placed in a foster family with people who care about him and whom he quite eagerly starts to care about. Now he can accept people being kind to him. And, of course, he receives the symbolically charged acne treatment kit. His scars are slowly healing.

Being part of a family and starting to use his real name indicate that Zits is no longer an outcast; he enters into affiliational relationships: »I need as many fathers as possible« (2007: 176), he admits. The need for his Indian father withers somehow, because now he has a better understanding of his Indian heritage. Zits needed his father not only as a child needs a parent, but also as a necessary link to his ethnic roots. The absence of his father therefore alludes to his disconnection from any Indian community. But, through his time travels, Zits acquires intimate knowledge about the Indian life and, more importantly, he connects with other Indians, so he no longer needs his father to teach him what it means to be an Indian.

Zits allows society to ›colonize‹ him, though it should be mentioned that he chooses the terms for his colonization: he becomes socially integrated, but his mind is no longer colonized by a self-deprecating image of Native Americans imbued in the dominant discourse. The fact that he stays with a white family, even though he still considers himself Indian, stresses that race and ethnicity are just dry concepts and real people are constantly negotiating their positions and are capable of love and forgiveness. The underpinning message of the novel is that any form of violence is nonsensical and no Justice can justify murder.

Tellingly, Zits/Michael is constructed as the embodiment of conflict negotiation between whites and Indians. As a mixed-blood, he occupies a potentially powerful and disruptive position. Half-breed identity issues resurface in the work of many contemporary Native American writers, from N. Scott Momaday to James Welch, Leslie Marmon Silko, Linda Hogan, Louise Erdrich, Louis Owens, and Gerald Vizenor. Whereas for most native writers, half-breeds find resolution in choosing to live in one of the two irreconcilable worlds, usually moving back to the tribal land, for Vizenor, mixed-bloods (or ›crossbloods‹, in his terminology) encapsulate the desire to hold on to a perpetual state of ambiguity and marginality that allows them to play the trickster part in white society in order to deconstruct the false images society has of them. However, unlike Vizenor, Sherman Alexie endows his mixed-bloods with a unique ability to reconcile the conflictive sides, ethnic harmony and empathy becoming a possible replacement for the hate and frustration that keep them segregated.

In *Flight*, two oppositional worlds are united in one character, similar to N. Scott Momaday's designing of Abel, the main character of *House Made of Dawn* (1968). But whereas Abel finds his way back to his native roots and distances himself from the harmful white world, disavowing whiteness and deeming the Indian and the white worlds irreconcilable opposites, Zits reaches a more generic understanding of human nature and takes his first step toward reconciliation by choosing to live with a white foster family, because he feels he could have a home irrespective of skin color. Ethnic grounds for choosing sides are discarded as irrelevant and defective in everyday life, for identity is fluid, constantly renegotiated and never really pinpointed. After Zits has been reborn as Michael, his foster mother remarks, »In a few months you'll be brand new« (2007: 180) vis-à-vis the effects of his acne treatment, which implies not only healing, but also clearing space for growth.

It should be highlighted that Sherman Alexie metaphorically delineates Crazy Horse as a mixed-blood, too. Crazy Horse has featured prominently as the archetypal hero in Alexie's fiction ever since his first book was published. A role model for most of his troubled characters, Crazy Horse is regarded as a source of strength for contemporary Native Americans, a vivid memento that they have not been completely defeated yet. In *Flight*, Alexie conveys the idea that mixed-bloods can be seen as the epitome of a bridge between the two conflicting worlds of natives and whites. Also, a very subtle paradigmatic shift in Alexie's ideology emerges in *Flight*, for the ethnic paradigm fades into the background, leaving room for a more integrative understanding of human nature and its flaws; a generic humanity that levels the field of ethnic tension and advocates for ethnic reconciliation.

Sherman Alexie's fiction prior to *Ten Little Indians* (2003) was ideologically structured around subverting and revising dominant discourses, demythologizing white stereotypes, offering native individuals a sense of community through storytelling, as well as questioning native authenticity and tradition. Alexie's 2003 short story collection, while denunciating racial and social injustice perpetuated within a multicultural America, also advocates »a common space for shared humanity« (Ladino 2009: 39) where identities are continually renegotiated according to variables such as race, culture, gender, or sexuality. As Åse Nygren notices in her interview with the native writer, Sherman Alexie's writing »shifted in emphasis from angry protests to evocations of love and empathy« (2005: 151).

Alexie's intention to distance his writing from a native perspective and to integrate his work into a larger, more encompassing frame of universality becomes obvious in *Flight*. This mutation is somehow surprising for a writer whose previous work was deeply rooted in a particular ethnicity that shaped and defined him both as an artist and as a human being. While it may be even more astonishing for readers familiar with Alexie's militant writing to read this plea for ethnic reconciliation, this move is not entirely inconceivable. In the 2003 short story collection *Ten Little Indians*, most native characters (who no longer live on the reservation) strive to reconcile the two opposite worlds (their native background and the predominantly white city) and are sometimes partially successful, but at other times fail miserably. Focusing on urban Indians, most of them with successful careers, the stories theorize modalities of negotiating identities in the contemporary multicultural American city, as Jennifer Ladino suggests in her article »A Limited Range of Motion«. There is also a sense of cross-ethnic solidarity in face of adversity, especially 9/11. Harmonizing the two worlds proves to be a difficult task for many native characters who choose to live some version of the American Dream in the white world, even though for most of them, it means cutting ties with their native kin. Nevertheless, *Flight* examines another issue connected to belonging: it is not the voluntary estrangement from the community or the relocation from the reservation to the big city (the personal choice of cutting ties), but the forced separation from the community. Zits' situation is a particularly odd one compared to Alexie's other characters: Since he has never lived in a native community, he does not experience the torments of Alexie's fictional characters on or off the reservation (even if they leave, the reservation still holds a significant place in their minds). ›Integration‹ (read: assimilation) into the mainstream society proves not too difficult for him. In fact, Zits is located similarly to John Smith in *Indian Killer*. Being completely uprooted from their communities, both characters struggle to find their ways back in any possible way. And since the only information available to them is circulated through mainstream discourse, they try to harmonize the image of the ›authentic Indian‹ with their lives. But the prescriptive Indianness disseminated in the media makes it next to impossible to translate into everyday living. John Smith commits suicide. As for Zits, by getting in contact with ›real Indians‹ (even though this authenticity is defined by the dominant discourse in the media), he manages to somehow correct this Vanishing Indian discourse. Their desire is the same —to be an ›authentic Indian‹, but for Zits this is merely a means to get closer to his absent father, to have some sense of family and community. At the end of the day, human communion proves far more important than any ethnic delineation in *Flight*, which thus pictures a post-race utopia in which the boundaries between ethnicities are broken down in favor of human spiritual unity.

NOTES

1 | Work on this chapter was supported by the European Social Fund in Romania, under the responsibility of the Managing Authority for the Sectoral Operational Programme for Human Resources Development 2007-2013 [Grant POSDRU/88/1.5/S/47646].
2 | Fanon's goal was to demonstrate that »what [was] often called the black soul [was] a white man's artifact« (1952/2008: 6). The same concept can be applied to ›the Indian ways‹.
3 | This process was extensively analyzed by Philip J. Deloria in his seminal study *Playing Indian* (1998). In the book, Deloria demonstrates how American society appropriated various aspects of native ethnic specificity in order to define themselves against European society, on the one hand, and to legitimize its presence on the continent, on the other hand. Americanness therefore incorporates Indianness, even if this acknowledged incorporation is only temporary.
4 | Representations of Native Americans have been constantly distorted in Euro-American discourses since the first contact (see, for instance, Christopher Columbus' letters after his ›discovery‹ of the ›West Indies‹) up to contemporary Hollywood blockbusters (like *Avatar* [2009]). The tendency to fictionalize and stereotype indigenous peoples in written texts and, more recently, in motion pictures, stems from the dominant group's need to justify past and present actions against the native population. Thus, the imaginary Indian gradually gained prominence over accurate accounts of native people or over self-representations. The Euro-American discourse, managed to carve a certain representation of the natives in the collective consciousness that influenced both the non-native and the native communities. See Berkhofer (1978), Slotkin (1973), Friar/Friar (1972), Churchill (1992), and Buscombe (2006) for only a handful of the most important studies on this topic.
5 | In *The World, the Text and the Critic*, Said initially employs these concepts for describing oppositional modalities in which critics engage with culture (filial relationships vis-à-vis the inherited culture and affiliation in relation to the adopted culture). Said argues that »if a filial relationship was held together by natural bonds and natural forms of authority—involving obedience, fear, love, respect, and instinctual conflict—the new affiliative relationship changes these bonds into what seem to be transpersonal forms – such as guild consciousness, consensus, collegiality, professional respect, class, and the hegemony of a dominant culture« (1983: 20).

REFERENCES

Alexie, Sherman (1996): *Indian Killer*. New York: Atlantic Monthly Press.
——— (2003): *Ten Little Indians*. New York: Grove Press.
——— (2007): *Flight*. New York: Black Cat.

Berkhofer, Robert F. (1978): *The White Man's Indian: Images of the American Indian from Columbus to the Present*. New York: Random House.

Bhabha, Homi K. (1994): *The Location of Culture*. London: Routledge.

Buscombe, Edward (2006): *»Injuns!«: Native Americans in the Movies*. London: Reaktion Books.

Churchill, Ward (1992): *Fantasies of the Master Race: Literature, Cinema, and the Colonization of American Indians*. Monroe: Common Courage Press.

Cohen, William B. (1970): »The Colonized as Child: British and French Colonial Rule«, *African Historical Studies* 3 (2), pp. 427–431.

Cox, James (1997): »Muting White Noise: The Subversion of Popular Culture Narratives of Conquest in Sherman Alexie's Fiction«, *Studies in American Indian Literatures* 9 (4), pp. 52–70.

Curtis, Lewis Perry (1968): *Anglo-Saxons and Celts: A Study of Anti-Irish Prejudice in Victorian England*. New York: New York University Press.

Deloria, Philip J. (1998): *Playing Indian*. New Haven: Yale University Press.

Fanon, Frantz (1952/2008): *Black Skin, White Masks*. London: Pluto Press.

Friar, Ralph E./Friar Natasha A. (1972): *The Only Good Indian: The Hollywood Gospel*. New York: Drama Book Specialists.

Huhndorf, Shari M. (2001): *Going Native: Indians in the American Cultural Imagination*. Ithaca: Cornell University Press.

Ladino, Jennifer K. (2009): »›A Limited Range of Motion?‹: Multiculturalism, ›Human Questions‹, and Urban Indian Identity in Sherman Alexie's *Ten Little Indians*«, *American Indian Quarterly* 21 (3), pp. 36–57.

Nygren, Åse (2005): »A World of Story-Smoke: A Conversation with Sherman Alexie«, *MELUS* 30 (4), pp. 149–169

Prashad, Vijay (2001): *Everybody Was Kung Fu Fighting: Afro-Asian Connections and the Myth of Cultural Purity*. Boston: Beacon Press.

Prucha, F. P. (1969). »Andrew Jackson's Indian Policy: A Reassessment«, *Journal of American History* 56 (3), pp. 527—539.

Said, Edward (1983): *The World, the Text and the Critic*. Cambridge: Harvard University Press.

Slotkin, Richard (1973): *Regeneration through Violence: The Mythology of the American Frontier, 1600–1860*. Middletown: Wesleyan University Press.

Sugano, Motoko (2005): »Inheritance and Expectations: The Ambivalence of the Colonial Orphan Figure in Post-Colonial Re-Writings of Charles Dickens's *Great Expectations*«, M.A. Thesis. Sydney: University of New South Wales.

Weich, David (2007): »Revising Sherman Alexie«, *Powell's Books* [online], 15 May, http://www.powells.com/interviews/shermanalexie.html. 25 October 2009.

This Space Called Science:
Spatial Approaches, Border Negotiations, and the Revision of Cultural Maps in Contemporary Popular Culture

JUDITH KOHLENBERGER

>»And the Geek Shall Inherit the Earth.«
A POPULAR T-SHIRT SLOGAN FOUND ON AMERICAN COLLEGE CAMPUSES

»Our society has undergone a paradigm shift. In the information age, you and I are the alpha males,« Dr. Leonard Hofstadter, experimental physicist and protagonist of the American hit sitcom *The Big Bang Theory* (CBS, 2007–present), assures himself and his fellow scientists during a fancy-dress party. The success of the show proves him right: ›Smart Is The New Sexy‹ was the tagline soon adopted by media all over the world. As if determined to prove the point, innovative infotainment shows, scientists as cult TV stars, and traditional TV formats which have become invested in science draw an unmatched number of viewers.[1] The times when science and its devotees used to be represented by one likable, yet hopelessly pathetic geek character seem to be long gone. Contemporary popular culture appears to wholeheartedly embrace what the world of science has to offer—and vice versa. Individual disciplines have started to appropriate and make avid use of popular media. News of the first proton collision at CERN, one of the most prestigious scientific endeavors of our time, were spread on *Twitter*, and there is hardly a budding scientist who does not present her findings on an online blog, thus making her research appear exciting, hip, and zeitgeisty. Science, it appears, has eventually left its dim laboratories and dusty shelves for good.

This chapter is set to explore contemporary spatial imagery in the conceptualization of science, which shall be linked to its heightened extent of amalgamation with the practices and styles of American popular culture. In accordance with sociological and cultural studies scholarship on boundary work and cultural map-making, respectively, science will be understood as a space whose borders are increasingly infiltrated by and mingled with popular culture. Conversely, it is the realm of popular culture, above all the channels of film, television, and new media, which readily exploits the literally »awesome authority that science possesses« (Broman 1998: 143) as a particularly feasible way of drawing audiences. This constellation will be examined in relation to the ubiquitous notion of coolness, which, after having conquered the realms of mass entertainment, advertisement, and fashion, now also invades the world of hard science: In the wake of augmenting the penetrability of its boundaries, science has finally discovered the formula for cool. It is, as I will argue in the following, due to an evident, novel emphasis on the ephemeral quality of coolness in recent representations of science that its territory is newly demarcated and its maps redrawn. Before, however, elaborating on the particularities of border negotiations between science and popular culture as allegedly distinct areas of American society, I shall first drop some science[2] on current maps and competing academic frameworks of scientific boundaries, to which this chapter aims to respond.

Border Patrols: Legitimation Discourses, Technoscience, and the American Knowledge Society

In his paradigmatic—and by now notorious—*The Postmodern Condition* (1979), Jean-François Lyotard famously characterized the widespread legitimation crisis of science as a defining feature of the postmodern age. According to Lyotard, the fact that we have become incredulous toward »the metanarrative apparatus of legitimation« and »no longer expect salvation to rise from [it]« (1979/1984: xxiv) constitutes a, or rather *the*, defining feature of postmodernity. The superseded metanarratives in question include, above all, the (modernist) notion of science as a means of progression and liberation, which is grounded in humanist thought and the French Revolution, and the Hegelian tradition of science for science's sake, in which »knowledge first finds legitimacy in itself« (Lyotard 1979/1984: 34). Lyotard's *Report on Knowledge*, the subtitle of his study, is widely understood as a classic theorization of contemporary society, on par with Henri Lefebvre's ideas on controlled consumption (1971), Guy Debord's ›society of spectacle‹ (1967) and Daniel Bell's ›postindustrial society‹ (1973). Additionally, Lyotard's account also constitutes, as Fredric Jameson has argued, »a thinly yeiled [sic] polemic« (1984: vii) against Jürgen Habermas' concept of *Legitimation Crisis* (1973) on the universal collapse of governing institutions. In contrast to Habermas, who saw the present period as a continuation of the yet incomplete project of Enlightenment, Lyotard's conceptualization defines postmodernity in radical opposition to the ethics, aesthetics, and epistemologies of modernism.[3]

As Paul Foreman explicates in his 2007 essay »The Primacy of Science in Modernity, of Technology in Postmodernity, and of Ideology in the History of Technology,« the modernist period can aptly be characterized by the messianic role it attributed to science. At the beginning of the twentieth century, the prime legitimation for science was its perceived usefulness and efficiency, two central criteria which all aspects of life had to measure up to. Social, individual, and national improvement constituted the era's top priorities,[4] which notably influenced the way science was perceived, legitimized, and interacted with in American public discourse. Not by chance, the modernist era witnessed the Progressive Movement, which led to a variety of fundamental reforms,[5] and according movements in the fine arts, such as precisionist painting and the growing popularization of photography and motion pictures for supposedly truthful and efficient representations of reality (Curtis 2009: 85). The ›Jazz Age‹, as the era has often, and sometimes rather derogatively, been dubbed,

> included radios that pulled human voices from the air, skyscrapers that emerged from city skylines, and advances in flight that promised to make aircraft nearly as commonplace as automobiles. Science sometimes became an important means of making sense of and successfully navigating the new world Americans saw emerging around them. Some scientists, for instance, portrayed science and scientific thinking as important aspects of American democracy and bulwarks against the spread of totalitarianism. (Thurs 2007: 95)

Science as the alleged epitome of meticulousness, efficiency, and progress not only perfectly accomplished the era's central promises, but epitomized its very ideals. The high status of science was further enhanced by the enormous popularity of individual researchers, above all Albert Einstein, who not only became a veritable media celebrity after his emigration to the United States, but who continued to shape the dominant image of the scientist for decades to come. Contemporary and shortly subsequent publications, such as Vannevar Bush's report to the U.S. government suggestively entitled *Science: The Endless Frontier* (1945), unmistakably reflect the high hope set in scientific discoveries and the prevailing image of scientists as new national heroes.[6]

This image had to endure noticeable modification with the advent of the Second World War and its atrocities, which could only be realized through several vital ›breakthroughs‹ in the fields of applied physics and biochemistry. First and foremost, this involves the dropping of the first atomic bomb, which did irreversible damage to the hitherto unblemished reputation of science. In the United States, this was even enhanced by the prominence and public demise of J. Robert Oppenheimer, the notorious ›father of the atomic bomb‹, who came to personify, as Bell evocatively phrased it, »the Janus-faced symbol of science as creator and destroyer« (1973: 398). New fields of research, such as genetics, surveillance technology, and nuclear energy, additionally emphasized the destructive rather than beneficial potentials dormant in scientific research. Eventually, this led to a thorough skepticism toward the utopian promises of science so readily believed in during earlier decades.

Furthermore, science did not only yield some rather unwelcome, or at the very least highly problematic, outcomes, it additionally also failed, as Marcel LaFollette points out, to live up to the high expectations it had itself raised: From the 1950s onwards, both academic and popular magazines recorded »a litany of promises that science would cure every disease, fix every problem, and brush away every fear—if only research was kept free of undue restraint. This ›song of science‹ [...] showed a public image of science that was not only inconsistent with reality but also politically unstable« (1990: 162). Public skepticism rose in direct relation to the increase in the political relevance of and funding for U.S. science and related sectors since the modernist period. Derek de Solla Price (1963) famously summarized the growing number of extensive, international science projects starting in the 1960s as ›Big Science‹.[7] His theory of uncontrollable and thus eventually detrimental growth of science incited a heated debate on science funding, risk management, and research ethics. Calls for an informed »knowledge *about* science that would underpin rational policy decisions on its finance and development« (Edge 1995: 6–7; my emphasis) were voiced, which marks the birth of modern science studies.

As reports like Lyotard's illustrate, the academic attention accorded to processes of epistemological production and scientific legitimation has, indeed, increased exponentially in the last few decades, which also stems from the fact that »[t]he technological, scientific, economic, and societal forces can hardly be segregated« (Weber 2003: 121)[8] at the beginning of the twenty-first century. The »seamless web« (Hughes 1986: 281) of science, technology, and society is generally summarized by the buzzword ›technoscience‹, which foregrounds the changed character of scientific practice, more than ever before dependent on high technology gadgets and expensive equipment.[9] Technoscience, a term attributed to Belgian philosopher Gilbert Hottois, has variously been theorized as a defining feature of Western culture since the 1970s.[10] In a similar vein like Lyotard, Bell in his seminal *The Coming of Post-Industrial Society* notes that a decisive property of this »interstitial time« (1973: 37) and thus a central argument in favor of the radical break theory is the heightened and permanently increasing importance of science. Bell argues

that the major source of structural change in society [...] is the change in the character of knowledge: the exponential growth and branching of science, the rise of a new intellectual technology, the creation of systematic research through R & D budgets, and, as the calyx of all this, the codification of theoretical knowledge. (1973: 44)

After a lengthy analysis of the precise parameters of this change, Bell emphatically concludes: »[T]he ethos of science is the emerging ethos of post-industrial society« (1973: 386). Expressions like knowledge society, information society, or professional(ized) society are thus used interchangeably with postmodernity, consumer and media society, or the late twentieth century. All of these terms attribute crucial importance to the role of codified and theoretical knowledge in daily life, as Americans are presently confronted with a »society based on the penetration of all its spheres of life by scientific knowledge« (Boehme/Stehr 1986: 8).

TRENCH WARFARE: MAPPING SCIENCE AND SCIENCE STUDIES

The enumeration of the central segments of American culture mentioned above—science, technology, and society—suggests three neatly segregated areas of activity. Accounts on the changed nature of present-day technoscientifc culture frequently seem to imply that it is only with the advent of Big Science projects that these three carefully bounded territories have become mingled and increasingly blur into each other. As scholarship in the sociology of science has, however, convincingly revealed, the boundaries of science to related segments of society (be it technology, medicine, or the humanities) have always been fuzzy and fiercely embattled by in- and outsiders. The conceptualization of professionalized knowledge via spatial imagery is, however, far from innovative. Famously, already Michel Foucault emphasized the fundamental spatiality of all knowledge since the turn of the last century:

[T]he formation of discourses and the genealogy of knowledge need to be analyzed, not in terms of types of consciousness, modes of perception and forms of ideology, but in terms of tactics and strategies of power. Tactics and strategies deployed through implantations, distributions, demarcations, control of territories and organizations of domains which could well make up a sort of geopolitics [...]. (1980: 77)

Inherent in Foucault's assertion and his admitted »spatial obsessions« (1980: 69) is the recognition that all cultural spaces, be they real or fictional, are subject to potentially violent border negotiations, which regulate the attribution of meaning, truth, and power. »[S]pace,« as Foucault maintains, »is fundamental in any exercise of power« (1991: 252).

It is hence not surprising that science, as one of the most authoritative and regulative segments of contemporary American society, should have been theorized in terms of space. Since the early 1980s, the demarcation of an area of professional activity as ›science‹ by means of ›boundary work‹ has gained a pivotal role in the study of science. Specifically, the term ›boundary work‹ refers to

»the discursive attribution of selected qualities to scientists, scientific methods, and scientific claims for the purpose of drawing a rhetorical boundary between science and some less authoritative residual non-science« (Gieryn 1999: 4–5).[11] Boundaries between science and related segments of society, between scientists and amateurs, and between scientific and popular productions, discoveries, and publications are accordingly understood as mere social conventions, continuously subject to revision, adaptation, and change. Famous historical cases of boundary work, such as the dispute between Thomas Hobbes and Robert Boyle (cf. Shapin/Schaffer 1985), illustrate that much is at stake in delineating science from pseudo-science: Only the former will be attributed cognitive and cultural authority, financial resources, and public prestige. Demarcation criteria are historically, socially, and culturally contingent, rather than universal and intuitive. Thomas Gieryn accordingly suggests regarding science as an empty space, »a space waiting for edging and filling« (1995: 407) by those personally interested in charging it with their own desires, hopes, and motivations.

Disciplinarily speaking, this approach is grounded in the sociology of science and the accordant social constructionist view of science. The latter postulates that scientific facts always emerge from already preexisting social relations and are thus invested with the preferences, motives, and epistemological preconditions of a particular social group (cf. Fisher 1990; Collins 1988). Rooted in Marx's and Durkheim's seminal works, social-constructivist studies strive to provide causal explanations for the particular outcomes of scientific practice by exploring »the momento-moment activities of scientists as they go about producing and reproducing scientific culture« (Restivo 1995: 107). The inescapable emphasis on the pivotal role of the scientist already suggests that sociological approaches, including the admittedly elucidative studies on boundary work, tend to center their analysis on what literary criticism might describe as the ›author‹ of a scientific fact, that is, the respective research conglomerate and their laboratories. Not surprisingly, thus, criticism on social constructivist approaches looms large on the postmodern horizon.

Recent scholarship in the study of science, generally subsumed under the umbrella term ›cultural studies of science‹, strives to problematize and eventually transcend the binary and only seemingly neat distinction between essentialist and social constructionist approaches to science. As Joseph Rouse explicates in his paradigmatic essay »What Are Cultural Studies of Science?« (1993):

[C]ultural studies of scientific knowledge take as their object of investigation the traffic between the establishment of knowledge and those cultural practices and formations which philosophers of science have often regarded as ›external‹ to knowledge. [...] Cultural Studies do not try to replace internalist accounts of knowledge by relying upon a privileged alternative explanatory framework (e.g. social factors), but neither do they grant epistemic autonomy to what is currently accepted as scientific knowledge. (1993: 4)

Bruno Latour's influential explorations of the construction of the laboratory and its objects (1983; 1987; 1993) and Donna Haraway's analysis of the metaphors informing the study of primatology (1989) are famous examples of this heterogeneous academic discipline combining philosophy, literary criticism, gender studies, sociology, and history.[12] In contrast to the more empirically oriented sociology of science, cultural studies recognize their own epistemic engagement in processes of legitimation, identity formation, and (dis)empowerment. The examples thus analyzed traditionally stem from what would currently constitute the periphery of established science, such as advertisements in the prestigious *Science* magazine (Haraway 1992). In accordance with the constructionist approach in cultural theory, scientific productions, be they publications, research projects, or the laboratory itself, are understood as texts in the broadest sense, that is, sites of meaning production and negotiation, and are accordingly subjected to close readings. This entails that visual and verbal re-presentations of science in a myriad of forms are scrutinized for their potential to produce, rather than merely reflect, knowledge through language, image, and discourse. Exceeding the realm of mere reflection, such cultural formations of science thereby actively contribute to a specific construction of both dominant and residual societal ideas. In contrast to sociological scholarship, the goal of such an approach is anything but the causal explanation of the epistemological consequences of the representations under scrutiny: Postcolonial theory and its cautioning against the non-reciprocal imposition of categories on an unwilling Other has demonstrated that this stance ultimately »reifies the boundaries between the interpretation and what it interprets« (Rouse 1993: 9).

With regard to the conceptualization of science as an entity with a significant spatial dimension, this entails that cultural studies not only realize their own epistemic and political investment in scientific production, but are well aware of how their own productions will, *nolens volens*, constitute yet another (more or less authoritative) specimen of boundary work. Remaining skeptical toward the proclaimed disinterestedness and objectivity of sociological accounts, cultural studies scholarship regularly sets its own, unavoidably subjective borders between science and related segments of professional activity. Gieryn's paradigmatic slogan—»[e]ssentialists *do* boundary work; constructivists *watch* it get done by people in society« (1995: 394)—might sound promising, but must eventually be exposed as inherently flawed. Indeed, by postulating what both essentialists and constructionists, scientists and non-scientists do or don't, Gieryn himself draws an elaborate map of the scientific domain. It is with respect to this inevitable bias that Pierre Bourdieu has ultimately described the scientific field as a battle ground in which supposedly passive observers are unwillingly engaged, as they »claim to impose the legitimate definition of the most legitimate form of science, i.e. natural science, in the name of epistemology or the sociology of science, [...] [for] the definition of the principles of evaluation of their own practice« (1975/1999: 34). From a cultural studies point of view, Bourdieu seems to imply that every researcher will eventually end up with their very own definition of science serving the specific purposes of their study.

Keeping this postulate in mind and retaining a basic »sensitivity to differences and contested meanings and identities« (Rouse 1993: 6) as a central premise for a cultural studies critique of science, it might, nonetheless, prove profitable to theorize contemporary interactions between popular culture and science in terms of spatial negotiation. As Gieryn suggests with reference to Clifford Geertz's concept of ›culturescape‹ (1973: 21), one must thus comprehend space as a particular *cultural*, rather than social, entity, which is much more contingent on its delineation from other territories, such a politics, religion, or pseudo-science, than on its classification by particular actors (1995: 415–416). This also helps to explain why some areas (›the periphery‹) of the scientific space are more heavily contested than others (›the core‹). Certain qualities, such as the Western ideals of objectivity and intellectual freedom, are solidified via frequent repetition and a large degree of consensus and thereby less frequently subjected to processes of negotiation and boundary work than more unstable and hazy properties of the scientific field (cf. Giddens 1979: 64; Gieryn 1995: 420). The same holds true for particular disciplines, research methods, national research traditions and funding, as well as academic writing styles. Indeed, feminist studies have demonstrated that the degree of border contestation is also highly dependent on gender, race, and class: Women engaged in the very same activities as their male colleagues have repeatedly been denied the professional title of trained expert.[13]

Rather than focusing on the social construction of scientific facts, it might hence prove more feasible to think of map-drawing in terms of the performative: It is through the »stylized repetition of acts« (Butler 1988: 519) that specific maps gain a certain degree of stability and appear incontestable and intuitive, rather than heavily combated, hazy, and unstable. As with the production of gendered bodies through performative acts, the repetition and control of borders (and adequate punishment in case of noncompliance or transgression) aids to solidify them and thereby defines a certain cognitive area as the ›natural‹ territory of science. Eventually, this will create a profound »appearance of substance« (1988: 520), as Judith Butler has compellingly argued in quite a different context. The space we call science could hence be exposed as part of »prevalent and compelling social fictions [...] which, in reified form, appear as the natural configuration« (Butler 1988: 524). As Butler's theory on gender performativity has, however, also demonstrated, this act remains quite independent of the individual actor and is anything but a radically free choice, being both historically and culturally contingent. In this respect, the concept of the performative can meaningfully account for the great sedimentation, not to speak of naturalization, of vast areas of the scientific wilderness. Contemporary Western science is, per definition, opposed to both religion and politics, and no elaborate argumentation, nor even deliberation, is needed to expose ufologists and phrenologists as attention-seeking pseudo-scientists. These examples vividly demonstrate that although its boundaries might be prone to blur, the vast majority of our common cultural map of science is presented to us as highly inflexible and hardly subject to revision at all.

INVASIONS: THE CONQUEST OF COOL

Recent (visual, but also verbal) representations of science and its parameters suggest that its territory is almost notelessly expanded to what are typically perceived as channels of popular culture—be it the cyberspace of social networks, blogs, and online video portals or the world of film and television. A prime means of how this invasion is undertaken are recent representations of science in terms of coolness, which has successfully extended its sphere of influence way beyond the traditional realms of fashion, advertisement, and mass entertainment. Conversely, of course, it is the space of science which may be understood as expanding its territory via the strategic utilization and adaptation of popular cultural cool. In that sense, coolness can be regarded as a viable cultural strategy for perceptively revising the borderlines of traditional science by allowing a heightened degree of fusion with the modes, aesthetics, and techniques of contemporary popular culture.

From an attitude of the 1960s counter-culture into a principal aesthetic norm of mainstream society, cool has indeed witnessed a triumphant evolution and progressively infiltrated more and more aspects of society. As Annette Geiger, Gerald Schröder, and Änne Söll put it:

Cool as a cultural strategy touches upon the domain of the individual and the collective, the aesthetic and the mental, the social and the political, but also economic dimensions, i.e., media and market, as well as gender, nationality, and race. The concerned field of research could not be any more extensive. (2010: 7)[14]

The vastness of its scope stems from an underlying ontological ambiguity: On the one hand, cool has devolved into a universal, yet elaborate term of approval, while its etymological roots on the other hand still evoke far-reaching connotations of coldness and emotional distance. Indeed, the indefinability of cool appears to emerge from an underlying juxtaposition of contradictory ideas, resulting in paradoxically conflicting effects (cf. Rice 2003: 221–224). Despite, or rather because, of this structural obscurity, coolness as an attitude, a sensibility, an emotional style, a cultural strategy or a form of zeitgeist, whichever kind of ontological status one assigns to it, must be treated as one of the most pervading qualities of contemporary postmodern society. Ulf Poschardt, for instance, notes that cool perfectly captures »the interpersonal climate in modern mass societies« (1999: 9),[15] while the trio of Geiger, Schröder, and Söll agrees that »despite the vagueness of its definition, coolness must be regarded as a central category of the twentieth and twenty-first century, which has [...] significantly shaped the cultural self-conception« (2010: 7).[16] The notion of cool displays a noticeable proximity to the values of contemporary information society, which establishes it, according to Alan Liu, as »the cultural dominant of our time« (2004: 76). Expressed more evocatively, cool thereby represents »the techno-informatic vanishing point of contemporary aesthetics, psychology, morality, politics, spirituality, and everything. No more beauty, sublimity, tragedy, grace, or evil: only cool or not cool« (Liu 2004: 3). While coolness, in the age of globalization and informatization, has gained an almost global intelligibility, it still retains its distinctly American flavor, as Peter Stearns, among several others,[17] argues: »The concept is distinctly American, and it permeates every aspect of American culture. [...] [T]he idea of cool, in its many manifestations, has seized a central place in the American imagination« (1994: 1).

A particularly avid example of how coolness increasingly operates as a dominant sensibility within almost all segments of Western, but particularly U.S.-American society is the Fermi National Accelerator Laboratory, better known by its abbreviation Fermilab, near Chicago, Illinois. A prime representative of contemporary, though nationalized, Big Science, Fermilab specializes in particle physics research and hosts one of the world's largest and most powerful particle accelerators, second only to CERN's Large Hadron Collider. Similar to CERN, Fermilab makes eager and proficient use of social media by hosting its very own *Facebook* profile, *Twitter* feed, and *YouTube* channel. Apart from these popular culture appearances, the site regularly stages cultural events, including concerts by both classical and contemporary musicians, hosts a performing arts series and an art gallery, and has its very own wildlife resort (a herd of bison roaming the campus). Rather than capitalizing on public lectures and sound educational films only, Fermilab eagerly employs alternative channels for popularization, corporate image work, and scientific self-legitimatization. Its *YouTube* channel, for instance, features videos of the staff choir ›The Fermilab Singers‹ and a handicraft tutorial on how to make your very own jelly bean universe (starring a young and agile scientist in candy-striped apron). Its verbal presentation to the public looks equally prosaic and apparently aims to liken the mundane daily routines of its researchers to that of bad-ass rock stars: »Who needs sleep when you have physics?« @*Fermilab_Today* recently tweeted (2011). Fermilab thus seems to have successfully jumped on the bandwagon of CERN's LHC, whose popular cultural exploitation already long suggested that »particle physics was the coolest thing on the planet« (2010: par. 2), as Alok Jha, science correspondent at *The Guardian*, aptly remarked in a recent article.

The various forms of coolness these research conglomerates play upon are, indeed, highly manifest: Without doubt, the shared cultural image of cool rests upon associations with youth(fulness), subculturality, high-tech gadgetry, and, if nothing else, dark sunglasses as the ultimate insignia of any veritably detached persona (cf. Pountain/Robins 2000: 7). More significantly in this case, however, the effect of coolness is produced by two central qualities which all of the aforementioned productions share: contradiction and unknowability. As Jeff Rice argues with regard to electronic discourse, a cool effect is achieved by »creating associations and emotional responses out of the combination of unlike words and images« (2003: 223). According to this proposition, coolness stems from the combination of contradictory images, an »ambiguous sensibility« (Pountain/Robins 2000: 43), resulting in conflicting outcomes. In the case of many global science projects, it is the obvious clash of the sublime and the trivial, the cognitive and the aesthetic, the high and the low culture exponent that generates a paradoxical final product, presenting us with the juxtaposition of seemingly random, disconnected images, such as the colorful amalgamation of cycling junior researchers, space travel interior design, and a comic strip version of *The Origin of Species* in a video clip issued by CERN (2007). If one follows Rice's theory of cool contradiction, it is this simultaneous reification of (supposedly) mutually exclusive qualities which lies at the source of cool.

Additionally, it is the elusive »ethos of the unknown« (Liu 2004: 9) which enhances the coolness factor of extensive research sites such as Fermilab. In contrast to educational texts conventionally issued by such research institutions, many productions through which Fermilab becomes manifest in the public perception seem to be utterly devoid of (scientific or other) content. As the examples elicited above demonstrate, neither its tweets nor its videos or tutorials are exclusively conceived to inform, educate, or instruct. What is foregrounded by these popular cultural channels is the affective, rather than the informative level. This largely coincides with Liu's argumentation, who compellingly points out that »[c]ool is the aporia of information« (2004: 179): Cultural texts are at their coolest, he maintains, when no information is forthcoming, when cool is hence symbolically »committing acts of destruction against [...] the content, form, or control of information« (Liu 2004: 8). For forging affective relations based on sentiment and embodied experience rather than mediated (scientific) knowledge, the elicitation of coolness proves to be a most viable strategy.

The use of such cool aesthetics as illustrated in Fermilab's new media productions exemplarily reflects how in the postmodern age »aesthetic impulses have spilled out of the self-consciously defined sphere of art into the spheres of the cognitive and the scientific on the one hand and the practical or moral on the other« (1992: 3), as Patricia Waugh eloquently argues by resorting to yet another spatial metaphor. Equally, I would propose to add, the scientific can be regarded as extending its competence, scope, and forms of interaction to related segments of activity. By proficiently adopting a strategy firmly grounded in popular cultural expressions, be it the language of advertisement or youth subcultures, science revises its line between the orthodox and the impermissible. The abovementioned examples indeed suggest that the representation of professional activity in a cool fashion is no longer coded as eccentric or even unscientific. Regarding the visual outputs of many contemporary Big Science endeavors, this can even be carried further: Coolness in terms of contradiction and unknowability has become a crucial element of its mission and (desired) corporate identity. This expansion of the scientific space may be linked to wider epistemic concerns of the postmodern technoscientific knowledge society. In that sense, the ubiquitous, yet imprecise and certainly non-factual notion of cool can well be regarded as responding to and complementing traditional discourses of scientific legitimation, which Lyotard has famously characterized as entirely obsolete. Following Lyotard's argumentation, coolness as a governing principle of scientific rhetoric must be conceived of as an instance of small and fluctuating narratives substituting the *grands récits*, the ›metanarratives‹ (Lyotard 1979/1984), toward which we have grown thoroughly incredulous. In this vein, it is for the sake of its own legitimation and authorization that the space called science is forced to substantially revise its boundaries.

CEASEFIRE, OR: SPATIALITY UNBOUND

There is much left to explore in the irregular, yet increasing overlaps and interfaces between adjacent cultural spaces. In contrast to studies on the popularization of scientific knowledge (cf. Hilgartner 1990; Schirrmacher 2008), work on the mutual, rather than unilateral, exchange between popular cultural production and science is yet underrepresented. Although the sociological concept of boundary work has been revealed as problematic with regard to its supposedly objective, disengaged observation of the subject matter, it can nonetheless yield remarkable insights into the dynamics of the historical and social contingencies of cultural spaces. Drawing on recent contributions from a cultural studies perspective allows theorizing specific border negotiations more adequately and proficiently. As the section headers of this chapter may have illustrated, the imagery of warfare and borderland controversy can provide an appropriate jargon for conceptualizing the instability of cultural maps. The strategic production of coolness in a myriad of forms and representations constitutes one of several highly effective troublemakers and helps to illustrate how the realm of science, far from being intuitive and stable, must be continuously re-defined. In a broader context, finally, it is not only the spatial that can help us shape and augment our understanding of the scientific and the discourses surrounding it. Equally, the fluctuating maps of science may substantially contribute to problematize contemporary theorizations of spatiality, demarcation processes, and borderlands.

NOTES

1 | A case in point is CBS' crime drama series *CSI: Crime Scene Investigation* (2000-present), which was crowned the most watched television show in the world and won the International Television Audience Award for a Drama TV Series at the Annual Monte Carlo Television Festival 2011. According to CBS Studios International, *CSI* and its equally popular spin-offs *CSI: Miami* (2002-2012) and *CSI: New York* (2004-present) attract more than seventy million viewers worldwide.

2 | ›To drop science‹ refers to the transmission of important information. The expression can be found in recent rap lyrics (cf. Anon. 2006). This piece of jargon is further evidence of how science not only seems to have reached the remotest segments of society, but has also substantially enhanced its coolness factor in the wake of it. See also Thurs (2007: 4).

3 | For Lyotard, the totality of society promoted by Habermas and his philosophical forefathers Hegel and Lukács must invariably lead to conformism, terrorism, and a dystopian ideal of consensus. In contrast to Habermas and the Frankfurt School, Lyotard is understood as a proponent of the radical break theory, which postulates a caesura between modernism and postmodernism. For a closer analysis of the ideological debate between Lyotard and Habermas, see, for instance, Jameson (1984: vii-x).

4 | See also Nisbet (1980) on how progress became a central ideal in the United States of the 1910s-1920s.

5 | These include, among others, the introduction of women's suffrage, the passage of the Federal Trade Commission Act, and the temperance movement. See Thurs (2007: 98-99) for a close analysis of the Progressive Movement's effects on scientific practice.

6 | For more information on the political implications of this elevated status of science, see, for instance, Kuznick (1987).

7 | Examples of Big Science projects include, among others, CERN and the Manhattan State Project, which was dedicated to the construction of the nuclear bomb.

8 | »Die Momente des Technischen, Wissenschaftlichen, Ökonomischen und Gesellschaftlichen sind kaum noch voneinander zu trennen.« All translations mine.

9 | Additionally, the growing dependence of science on grants and international cooperation has had wide-reaching consequences for its relationship to politics, the financial sector, and the military. See also Weber (2003: 128) and Felt/Nowotny/Taschwer (1995: 53).

10 | Marcus (1995) and Ihde/Selinger (2003) offer good introductions to recent scholarship on technoscience.

11 | The concept of boundary work seems to emerge in Barnes (1974) and is more prominently elaborated by Gieryn (1983; 1995; 1999). Since then, a variety of (sociological) studies have been devoted to boundary work in general (Shapin 1982) or the empirical analysis of particular historical episodes (Shapin/Schaffer 1985; Jasanoff 1990).

12 | See the edited collection by Reid/Traweek (2000) for recent examples of cultural studies scholarship on science.

13 | Conversely, a female scientist might have frequently been denied the status of a real woman, as these categories were regarded as diametrically opposed. See, for instance, Schiebinger (1989), who perceptively shows how midwifery is relegated to the realms of tradition, prescience, or mere nursing. Nineteenth-century U.S.-American cookbooks and home economics manuals written by women have equally be dismissed as mere fancy rather than professional handicraft and expertise. For further insightful scholarship on feminism and science, see Haraway (1994), Fox Keller (1985), and Harding (1992).

14 | »Cool als Kulturtechnik berührt die Bereiche des Individuellen und Kollektiven, des Ästhetischen und des Psychischen, des Sozialen und Politischen, aber auch die Dimensionen der Ökonomie, d.h. der Medien und des Marktes, ebenso wie die der Geschlechter, Nationalitäten und Hautfarben. Das betroffene Forschungsfeld könnte also nicht weiter gefasst sein.« (All translations are mine.)

15 | »[...] das zwischenmenschliche Klima in modernen Massengesellschaften.«

16 | »Coolness muss trotz aller Vagheit in der Definition als eine zentrale Kategorie des 20. und 21. Jahrhunderts betrachtet werden, die das kulturelle Selbstverständnis [...] maßgeblich geprägt hat.«

17 | See also Joel Dinerstein's forthcoming monograph *The Mask of Cool: Jazz, Film Noir, and Existentialism in Postwar America* (University of Chicago Press), in which he delineates the origins of the concept of cool as it emerges in America's vibrant jazz culture.

References

Anon. (2006): »Drop Science, Drop Knowledge«, *The Rap Dictionary* [online], 19 December, http://www.rapdict.org/Drop_science,_drop_knowledge. 4 April 2011.

Barnes, Barry (1974): *Scientific Knowledge and Sociological Theory*. London: Routledge & Kegan Paul.

Bell, Daniel (1973): *The Coming of Post-Industrial Society*. New York: Basic.

Boehme, Gernot/Stehr, Nico (1986): *The Knowledge Society: The Growing Impact of Scientific Knowledge on Social Relations*. Dordrecht: Kluwer Academic Publishers.

Bourdieu, Pierre (1975/1999): »The Specificity of the Scientific Field and the Social Conditions of the Progress of Reason«, *Social Science Information* 14, pp. 19–47, rpt. in Mario Biagioli (ed.), *The Science Studies Reader*. New York: Routledge, pp. 31–50.

Broman, Thomas (1998): »The Habermasian Public Sphere and ›Science in the Enlightement‹«, *History of Science* 36, pp. 123–150.

Bush, Vannevar (1945): *Science—The Endless Frontier: A Report to the President on a Program for Postwar Scientific Research*. Washington: Government Printing Office.

Butler, Judith (1988): »Performative Acts and Gender Constitution: An Essay in Phenomenology and Feminist Theory«, *Theatre Journal* 40 (4), pp. 519–531.

CERN (2007): »Particle Hunters«, *CERNTV* [online], 2 November, http://www.youtube.com/watch?v=nifTTSfqubM. 18 May 2011.

Collins, Harry M. (1988): »Public Experiments and Displays of Virtuosity: The Core-Set Revisited«, *Social Studies of Science* 18 (4), pp. 725–748.

Curtis, Scott (2009): »Images of Efficiency: The Films of Frank B. Gilbreth«, in Vinzenz Hediger/Patrick Vonderau (eds.), *Films that Work: Industrial Film and the Productivity of Media*. Amsterdam: Amsterdam University Press, pp. 85–99.

De Solla Price, Derek (1963): *Little Science, Big Science*. New York: Columbia University Press.

Debord, Guy (1967/1995): *The Society of the Spectacle* (trans. Donald Nicholson-Smith). New York: Zone Books.

Edge, David (1995): »Reinventing the Wheel«, in Sheila Jasanoff/Gerald E. Markle/James C. Petersen/Trevor Pinch (eds.), *Handbook of Science and Technology Studies*. Thousand Oaks: Sage Publications, pp. 3–23.

Felt, Ulrike/Nowotny, Helga/Taschwer, Klaus (1995): *Wissenschaftsforschung: Eine Einführung*. Frankfurt: Campus Verlag.

Fermilab_Today (2011): »Who needs sleep when you have physics?«, @*FermilabToday* [online], 30 November, http://twitter.com/#!/FermilabToday/statuses/141761280565448704. 2 December 2011.

Fisher, Donald (1990): »Boundary Work and Science: The Relation between Power and Knowledge«, in Susan E. Cozzens/Thomas F. Gieryn (eds.), *Theories of Science in Society*. Bloomington: Indiana University Press, pp. 98–119.

Foreman, Paul (2007): »The Primacy of Science in Modernity, of Technology in Postmodernity, and of Ideology in the History of Technology«, *History and Technology* 23, pp. 1–152.

Foucault, Michel (1980): *Power/Knowledge: Selected Interviews and Other Writings 1972–1977*. New York: Pantheon Books.

―――― (1991): »Space, Knowledge, and Power«, in Paul Rabinow (ed.), *The Foucault Reader: An Introduction to Foucault's Thought*. London: Penguin Books, pp. 239–256.

Fox Keller, Evelyn (1985): *Reflections on Gender and Science*. New Haven: Yale University Press.

Geertz, Clifford (1973): *The Interpretation of Culture*. New York: Basic Books.

Geiger, Annette/ Schröder, Gerald/Söll, Änne (2010): »Coolness—Eine Kulturtechnik und ihr Forschungsfeld: Eine Einführung«, in Annette Geiger/Gerald Schröder/Änne Söll (eds.), *Coolness: Zur Ästhetik einer kulturellen Strategie und Attitüde*. Bielefeld: transcript Verlag, pp. 7–16.

Giddens, Anthony (1979): *Central Problems in Social Theory*. Berkeley: University of California Press.

Gieryn, Thomas F. (1983): »Boundary-Work and the Demarcation of Science from Non-Science: Strains and Interests in Professional Ideologies of Scientists«, *American Sociological Review* 48, pp. 781–795.

―――― (1995): »Boundaries of Science«, in Sheila Jasanoff/Gerald E. Markle/James C. Petersen/Trevor Pinch (eds.), *Handbook of Science and Technology Studies*. Thousand Oaks: Sage Publications, pp. 393–443.

―――― (1999): *Cultural Boundaries of Science: Credibility on the Line*. Chicago: University of Chicago Press.

Habermas, Jürgen (1973/1976): *Legitimation Crisis* (trans. Thomas McCarthy). London: Heinemann.

Haraway, Donna (1989): *Primate Visions: Gender, Race, and Nature in the World of Modern Science*. New York: Routledge.

―――― (1992): »The Promises of Monsters: A Regenerative Politics for Inappropriate/d Others«, in Lawrence Grossberg/Cary Nelson/Paula Treichler (eds.), *Cultural Studies*. New York: Routledge, pp. 295–337.

―――― (1994): »A Game of Cat's Cradle: Science Studies, Feminist Theory, Cultural Studies«, *Configurations* 2 (1), pp. 59–71.

Harding, Sandra (1992): *Whose Science? Whose Knowledge? Thinking from Women's Lives*. Ithaca: Cornell University Press.

Hilgartner, Stephen (1990): »The dominant view of popularization: Conceptual problems, political uses«, *Social Studies of Science* 20 (3), pp. 519–539.

Hughes, Thomas P. (1986): »The Seamless Web: Technology, Science, Etcetera, Etcetera«, *Social Studies of Science* 16, pp. 281–292.

Ihde, Don/Selinger, Evan (2003): *Chasing Technoscience: Matrix for Materiality*. Bloomington: Indiana University Press.

Jameson, Fredric (1984): »Foreword«, in Jean-François Lyotard, *The Postmodern Condition: A Report on Knowledge* (trans. Geoff Bennington/Brian Massumi). Manchester: Manchester University Press, pp. vii–xxv.

Jasanoff, Sheila (1990): *The Fifth Branch: Science Advisers as Policymakers*. Cambridge: Harvard University Press.

Jha, Alok, et al (2010): »How Science Became Cool«, *The Guardian* [online], 13 April, http://www.guardian.co.uk/science/2010/apr/13/science-cool. 28 October 2010.
Kuznick, Peter J. (1987): *Beyond the Laboratory: Scientists as Political Activists in 1930s America*. Chicago: University of Chicago Press.
LaFollette, Marcel C. (1990): *Making Science Our Own: Public Images of Science, 1910–1955*. Chicago: University of Chicago Press.
Latour, Bruno (1983/1999): »Give Me a Laboratory and I Will Raise the World«, in Karin Knorr-Cetina/Michael Mulkay (eds.), *Science Observed*. Thousand Oaks: Sage Publications, pp. 141–170, rpt. in Mario Biagioli (ed.), *The Science Studies Reader*. New York: Routledge, pp. 258–275.
─── (1987): *Science in Action: How to Follow Scientists and Engineers through Society*. Cambridge: Cambridge University Press.
─── (1993): *We Have Never Been Modern*. Cambridge: Harvard University Press.
Lefebvre, Henri (1971/2002): *Everyday Life in the Modern World*. London: Continuum Books.
Liu, Alan (2004): *The Laws of Cool: Knowledge Work and the Culture of Information*. Chicago: University of Chicago Press.
Lyotard, Jean-François (1979/1984): *The Postmodern Condition. A Report on Knowledge* (trans. Geoff Bennington/Brian Massumi). Manchester: Manchester University Press.
Marcus, George E. (1995): *Technoscientific Imaginaries: Conversations, Profiles and Memoirs*. Chicago: University of Chicago Press.
»The Middle-Earth Paradigm« (2007), *The Big Bang Theory*, Season 1, Episode 6 (CBS, writ. Dave Goetsch, David Litt, and Robert Cohen, dir. Mark Cendrowski).
Nisbet, Robert (1980): *History of the Idea of Progress*. New York: Basic Books.
Poschardt, Ulf (1999): *Cool*. Hamburg: Rogner & Bernhard Verlag.
Pountain, Dick/Robins, David (2000): *Cool Rules: Anatomy of an Attitude*. London: Reaktion Books.
Reid, Roddey/Traweek, Sharon (2000): *Doing Science and Culture*. New York: Routledge.
Restivo, Sal (1995): »The Theory Landscape in Science Studies: Sociological Traditions«, in Sheila Jasanoff/Gerald E. Markle/James C. Petersen/Trevor Pinch (eds.), *Handbook of Science and Technology Studies*. Thousand Oaks: Sage Publications, pp. 95–113.
Rice, Jeff (2003): »Writing about Cool: Teaching Hypertext as Juxtaposition«, *Computers and Composition: An International Journal for Teachers of Writing* 20 (3), pp. 221–236.
Rouse, Joseph (1993): »What are Cultural Studies of Scientific Knowledge?«, *Configurations* 1 (1), pp. 57–94.

Schiebinger, Londa (1989): *The Mind Has No Sex? Women in the Origins of Modern Science*. Cambridge: Harvard University Press.

Schirrmacher, Arne (2008): »Nach der Popularisierung: Zur Relation von Wissenschaft und Öffentlichkeit im 20. Jahrhundert«, *Geschichte und Gesellschaft* 34 (1), pp. 73–95.

Shapin, Steven (1982): »History of Science and its Sociological Reconstructions«, *History of Science* 20, pp. 157–211.

———/Schaffer, Simon (1985): *Leviathan and the Air Pump: Hobbes, Boyle and the Experimental Life*. Princeton: Princeton University Press.

Stearns, Peter N. (1994): *American Cool: Constructing a Twentieth-Century Emotional Style*. New York: New York University Press.

Thurs, Daniel (2007): *Science Talk: Changing Notions of Science in American Popular Culture*. New Brunswick: Rutgers University Press.

Waugh, Patricia (1992): *Practising Postmodernism, Reading Modernism*. London: Edward Arnold Publishers.

Weber, Jutta (2003): *Umkämpfte Bedeutungen: Naturkonzepte im Zeitalter der Technoscience*. Frankfurt: Campus Verlag.

Contributors

Diana Benea is a Ph.D. candidate in American literature at the University of Bucharest, where she is working on a thesis on Thomas Pynchon's later fiction. Diana has published on Thomas Pynchon, Herta Müller, and literary histories of Eastern Europe. Her research interests include contemporary American literature, critical theory, and Eastern European Studies.

Georg Drennig studied North American Studies at the University of Vienna, Austria, and Georgetown University, U.S., and is now a Ph.D. candidate in the Advanced Research in Urban Systems program at the University of Essen/Duisburg, working on environmental imaginaries of Vancouver. His main research interests are spatially-turned Cultural Studies and ecocriticism. Within the latter field, he has been called an adherent to the ›school of stone-kicking realists‹, a designation he is proud of.

Michael Fuchs is an adjunct professor in American Studies at the University of Graz, Austria. He has co-edited *ConFiguring America: Iconic Figures, Visuality, and the American Identity* (2013) and *Landscapes of Postmodernity: Concepts and Paradigms of Critical Theory* (2010). Currently, he is working on three projects—a revision of his dissertation on self-reflexivity in horror cinema, a project on the *Supernatural*verse, and a book on playing *with* video games.

Maria-Theresia Holub holds a Ph.D. in Comparative Literature from SUNY Binghamton and is currently employed as a research and teaching associate in the Department of American Studies, University of Graz, Austria. Her specialization lies in the fields of Postcolonial Studies, Border Studies, migrant literatures, and feminist theory. Maria-Theresia has published articles and presented papers on these and other issues at various international conferences.

Ida Jahr is currently a Ph.D. fellow in the John F. Kennedy Institute for North American Studies, Free University Berlin, Germany, and a board member of the American Studies Association Norway. She is writing her dissertation on the work of Sigmund Skard, the first professor of American literature in Norway, and his ambivalent relationship to the place which was both his object of study and his source of funding. Ida holds a B.A. in English and an M.A. in North American Area Studies from the University of Oslo, Norway, and has taught courses in English and American Studies at the University of Oslo, Oslo University College, and the Free University of Berlin.

Judith Kohlenberger holds an M.A. in English and American Studies from the University of Vienna, Austria. Her diploma thesis on coolness as a cultural strategy in contemporary U.S.-American cinema was awarded the 2010 Fulbright Award in American Studies and the 2010 Academic Excellence Award by the Austrian Federal Ministry of Science. In 2012, Judith was awarded a DOC fellowship from the Austrian Academy of Sciences for her Ph.D. thesis, which explores notions of coolness in recent representations of natural sciences within U.S.-American popular culture. Her research interests include cultural and literary theory, Gender Studies as well as hemispheric approaches to the Americas.

Leopold Lippert is a Ph.D. candidate in American Studies at the University of Vienna and is currently working as a research and teaching associate in the Department of American Studies of the University of Graz. His Ph.D. project, tentatively entitled »Performing America Abroad«, deals with the ›Americanness‹ of Austrian cultural and academic practice. He has published on transnational issues as well as queer theater and film and situates his research at the intersection of queer theory, Performance Studies, and transnational American Studies.

Evelyn P. Mayer is a Ph.D. candidate in American Studies at Johannes Gutenberg University Mainz (FTSK Germersheim), where she is working on a thesis on Canada–U.S. border fiction. She spent the 2008–2009 academic year at Carleton University (Ottawa, Ontario) and at the Border Policy Research Institute, Western Washington University (Bellingham, Washington). In addition to literary and border studies, her research interests include cultural geography and translation studies. Evelyn holds a master's degree in conference interpreting for German, English, and French and works as a senior lecturer at Saarland University in Saarbrücken.

Madalina Prodan recently earned a Ph.D. in Philology from the University of Iași, Romania, with a thesis on Sherman Alexie's fiction. She has published on the representation of Native Americans in film and Sherman Alexie's short fiction. In addition to Native American studies, her research interests include gender studies and film studies.

Jeff Thoss recently completed his Ph.D. thesis entitled »Metalepsis in Contemporary Popular Fiction, Film, and Comics« at the University of Graz and now teaches English at the Free University of Berlin. His research interests include narrative theory, intermediality, and popular culture, topics which he has also published on, most recently a contribution to the collection *Unnatural Narratives—Unnatural Narratology* (2011). Jeff's current project traces media rivalry through English literary history.

Julia van Lill (née Schnabler) earned a Bachelor of Arts (Hons) degree in English at Royal Holloway in London. After graduation, she moved to Wiener Neustadt, Austria, where she was instrumental in setting up the Business Consultancy International program at the University for Applied Sciences Wiener Neustadt. After three years in Austria, Julia returned to London to complete a Master's degree at King's College London. Currently, she is a Ph.D. candidate in English and American Studies at the University of Vienna and working part-time at the University of Applied Sciences Vienna.

Yvonne Völkl studied French philology at the University of Graz, Austria, the Université Sorbonne Nouvelle—Paris III, France, and the Université de Montréal, Canada. In 2011, she received a Ph.D. in French Studies from the University of Graz with a thesis on Jewish migrant literature in Quebec. She is currently teaching classes on Quebec, France, and other francophone cultures in the Department of Romance Languages at the University of Graz. Yvonne has co-edited *Cultural Constructions of Migration in Canada/Constructions culturelles de la migration au Canada* (2011).

Index

0-9

9/11 145, 146, 177, 184

A

Abu Ghraib 79
affiliation 180–12
aging 63
Ahab (character in *Moby-Dick*) 174
Alexie, Sherman 173–85
Allegheny Ridge Line 165
American Dream 10, 11, 109, 117, 124n24, 184
American Gothic 107–109, 115, 120n12, 120n13, 120n17, 121n19; *see also* Gothic
American Progress (painting) 51n7
American Renaissance 47
American Studies Association 19, 24, 25
Anderson, Benedict 10, 48, 117, 164
Anne books (series by L. M. Montgomery) 58, 60–66, 71
Anzaldúa, Gloria 11, 33n4, 137
Appadurai, Arjun 23, 25, 43–44
Atwood, Margaret 12, 144, 149—152
Awakening, The (novel) 61

B

Bakhtin, Mikhail M. 94
Barthes, Roland 107
Battle of Little Bighorn 181
Baudrillard, Jean 11, 107
Bell, Daniel 188, 190
Benesch, Klaus 10, 107
Benjamin, Walter 45
Bergland, Renée L. 164, 165
Berlant, Lauren 24, 33n13
Berlin Wall 42
Bhabha, Homi K. 20, 33n4, 43, 48, 178
Big Bang Theory, The 187
Bloom, Harold 120n18
Bon Cop, Bad Cop 133–42
border 11–12, 60–71, 131–42, 147—54, 191–3; *see also* boundary
Botting, Fred 113
boundary 92–99, 189–90, 198, 200
Bourdieu, Pierre 193, 194
Brontë, Charlotte 60
Bush, George W. 77
Butler, Judith 194

C

capitalism 48, 49, 117
Carlson, Marvin 34n14
Carroll, Lewis 57
Carter, Kevin (photographer) 120n14
CERN 196, 199

Chakrabarty, Dipesh 42–3
Chavez, Hugo 86
Chicago Grain Exchange 39, 41, 45, 47
Chopin, Kate 91
chronotope 94–5; see also timespace
City Upon a Hill 9, 105
Coleridge, Samuel Taylor 57, 70
Columbus, Christopher 185n4
Conquergood, Dwight 34n14
cool 195–197
Crazy Horse 183
CSI: Crime Scene Investigation 199n1
cultural memory 179
Curtis, Scott 189

D

Danielewski, Mark Z. 103–22
Debord, Guy 188
de Certeau, Michel 88n6
deconstructionism 121n23
Deloria, Philip J. 24, 33n8, 185n3
Derrida, Jacques 29, 122n25, 170n3; see also haunting; see also hauntology
de Solla Price, Derek 190
de Tocqueville, Alexis 40, 43, 50
Dinerstein, Joel 200n17
DMZ 75–88
Duhamel, Georges 40, 44, 48
Durkheim, Emile 192
dystopia 76–82, 88n2, 199n3

E

Einstein, Albert 189
Ellison, Ralph Waldo 45–6
Emily books (series by L. M. Montgomery) 58, 66–70
Empire State Building 81

Erdrich, Louise 182
eroticism 66, 70
escapism 99
European Association of American Studies 41

F

Facebook 196
family 113–115, 122n27, 180–2
Fanon, Frantz 176, 185n2
Fermilab 196
Fiedler, Leslie 120n17
filiation 180–2
Fishkin, Shelley Fisher 19, 24
Flight 173–85
Fluck, Winfried 46
Foucault, Michel 27, 75, 89n1, 118, 191
Founding Fathers 9, 83
Fourth of July 83
Freud, Sigmund 111, 114, 120n18, 122n25, 153
frontier myth 10, 11, 108, 110–3, 168, 189; see also Turner, Frederick Jackson
future 25, 39–50, 68, 72, 76, 81, 84, 85, 97, 108, 131, 165, 169–70

G

Gadsden flag 81
Gast, John 51n7
Gellner, Ernest 48
gender 112, 193, 194, 200n13
Genette, Gerard 93
ghost 32, 161–70; see also specter
Giddens, Anthony 194
Gieryn, Thomas F. 192, 193, 194–5, 200n11
Giles, Paul 33n4

Gleason, Philip 45–6
globalization 20–22, 48–49, 109, 152
Gordon, Avery F. 29, 162, 167, 169
Gothic 69, 107, 111
Gravity's Rainbow 169–70
Guantanamo 79
Guevara, Che 86

H

Habermas, Jürgen 188, 199n3
Halberstam, Judith 25, 33n12
Halloween 117, 122
Haraway, Donna 193, 200n13
Harvey, David 33n12
haunting 26–32, 108–10, 115, 162–6
hauntology 29, 122n25
Hayles, N. Katherine 120n12
Hegel, Georg Wilhelm Friedrich 199n3
heterotopia 75–6, 168
Hogan, Linda 182
Hollywood 117, 122, 176
home 59–72, 92, 110–6, 120, 150
House Made of Dawn 183
House of Leaves 103–22
Hutcheon, Linda 150
hybrid space 64; see also in-betweenness

I

imagined community 11, 117–8, 161; see also Anderson, Benedict; see also national identity
imperialism 10, 12, 108, 117
in-betweenness 61, 150, 178–9
Indian Killer 175, 184
Indian Removal Act, The 178
Irish American 179
Iser, Wolfgang 98

J

Jackson, Andrew 178
James, Henry 113
Jameson, Fredric 188, 199n3
Jamestown 110
Jane Eyre 60, 68
Jazz Age 189

K

Kaplan, Amy 20, 32n2
Kennedy, John F. 10
King, Martin Luther 11
King, Stephen 91, 94–9
Kooijman, Jaap 33n4

L

landscape 59–71
Latour, Bruno 193
Lefebvre, Henri 12, 188
Lipsitz, George 25, 33n12
Little Bighorn 176
Liu, Alan 196, 197
Los Angeles 91, 94–6
Lotman, Jurij 91–4
Lukács, György 199n3
Lyotard, Jean-François 188, 190, 198, 199n3

M

MacLennan, Hugh 130, 142
Maine 91
Malina, Debra 96
Manifest Destiny 43, 45, 110, 117
Mao, Tse-tung 86
Marx, Karl 192

Mason & Dixon 161–70
Mason-Dixon line 162
Mayflower 44
McHale, Brian 161
McKenzie, Jon 26
McLuhan, Marshall 151–2
Melville, Herman 174
metalepsis 91–100, 105, 119n6
Mitchell, Don 76, 88n1
Moby-Dick 174
Momaday, N. Scott 182, 193
Montgomery, Lucy Maud 57–72
myth 22, 47, 83, 110, 161, 168, 169

N

national identity 141, 147–8, 153; see also imagined community
National Symbolic 20
Native Americans 11, 21–31, 110, 161–70, 173–84
nature 58–71, 92, 164
New Orleans 79
nihilism 121n23
No Name City 22–3, 30–3
nostalgia 45, 65, 67

O

Obama, Barack 9, 81
Olson, Charles 105
Owens, Louis 182

P

Pan, Peter 63
Paxton Boys' Massacre 165, 169
Peace Arch 154
Pease, Donald E. 32n2

performativity 26–31, 51, 107, 194
Pilgrim, Billy 182
Pocahontas 110
popular culture 98–100, 154, 187–99
postcolonialism 107, 153, 177, 193
post-race age 184
power 20, 27, 29, 32, 33n5, 34n14, 42, 49, 61, 80, 84–8, 105, 107, 146–50, 166–8, 179–82, 191, 193
Pratt, Mary Louise 33n4
Pynchon, Thomas 161–70

R

race 11, 184; see also Native Americans
Radway, Janice 29
reality 63–5, 105, 189
repression 69, 114–8, 162–4, 178–9
Roach, Joseph 28, 34n14
Rouse, Joseph 192, 193, 194
Ryan, Marie-Laure 93

S

Said, Edward W. 10, 180, 185n5
Sedgwick, Eve Kosofsky 114
Shelley, Percy Bysshe 57
Siege of Fort Pitt 169
Silko, Leslie Marmon 182
Simpsons, The 101n5
Skard, Sigmund 39–52
Slaughterhouse-Five 186
slavery 43, 45, 108, 164
Slothrop, Tyrone 169
Slotkin, Richard 110, 185n4
Smith, John (character in Sherman Alexie's fiction) 175, 184
Smith, John (explorer) 110
social constructivism 192
Soja, Edward W. 12, 33, 88

Space Race 10, 105
specter 29, 110, 170; see also haunting
Stearns, Peter 196
stereotype 138, 154, 183
Surfacing 12, 148–50

T

Tea Party movement 9, 81
Third Space 20, 33n11
time 40, 42, 68, 108, 122n25, 175; see also future
timespace 43, 44, 68, 94–6, 105; see also chronotope
transnationalism 23–31, 48–9, 105, 152
trauma 11, 81, 137, 173–75, 180–2
traveling 44, 65–9, 176
trickster 34n14, 182
Trudeau, Pierre Elliott 152
Turner, Frederick Jackson 11, 45, 107, 110
Twin Towers 145
Twitter 187, 196

U

uncanny 69, 111, 119n7, 120n18, 122n25
unendliche Geschichte, Die (Neverending Story, The) 93–4
utopia 75–6, 112, 169, 184, 190

V

Vidler, Anthony 111
violence 180–1
Vizenor, Gerald 182

W

Waterloo (Austrian singer and performer) 21–2, 30–2, 33n7
War of Secession 83
War on Terror 10, 77, 85
Washington, George 164, 165
Waugh, Patricia 198
weapons of mass destruction 87
Weber, Max 44
Weinstock, Jeffrey Andrew 162
Welch, James 182
westward expansion 30, 45, 47, 51, 168; see also frontier myth; see also Manifest Destiny
Whitman, Walt 118
wilderness 10; see also nature
Wild West 21–2
Williams, Tony 114
Winthrop, John 10
World Trade Center 145; see also 9/11

Y

Young, Robert 42
YouTube 196

Z

Zinn, Howard 11
Žižek, Slavoj 119n11